Nutrient Cycling in
Tropical Forest Ecosystems

Nutrient Cycling in Tropical Forest Ecosystems

Principles and Their Application in Management and Conservation

Carl F. Jordan

Institute of Ecology, University of Georgia, Athens, Georgia 30602, USA

JOHN WILEY & SONS

Chichester · New York · Brisbane · Toronto · Singapore

Copyright © 1985 by John Wiley & Sons Ltd.

Library of Congress Cataloging in Publication Data

Jordan, Carl F.
  Nutrient cycling in tropical forest ecosystems.

  Bibliography: p.
  Includes index.
    1. Biogeochemical cycles—Tropics.   2. Forest
ecology—Tropics.   3. Forest management—Tropics.
4. Forest conservation—Tropics.   I. Title.
QH84.5.J67   1985   574.5'2642   85–6457

ISBN 0 471 90449 X

British Library Cataloguing in Publication Data

Jordan, Carl F.
  Nutrient cycling in tropical forest ecosystems :
  principles and their application in management
  and conservation.
  1. Forests and forestry—Tropics   2. Soil
  chemistry
  I. Title
  631.4'1   SD247

ISBN 0 471 90449 X

Printed and bound in Great Britain

Table of Contents

Preface

Tropical forests are being cleared at increasing rates. Many people are concerned, because they sense that the destruction represents an important loss. Tropical deforestation for agriculture and pasture usually represents a loss of the commercial value of the wood, because trunks are often burned in place after they are cut. Tropical deforestation also is important for non-commercial reasons. The forests of the tropics may influence climate, and the global balances of carbon and atmospheric pollutants. In addition, deforestation often results in increased erosion. This lowers the productive capacity of the soil, which already may be low because of the low native fertility of many tropical soils. Erosion also causes siltation in reservoirs that supply water and electricity to many large cities in the tropics.

There are other less obvious values of tropical forests which are lost when the forests are cut. The forests are a home to an incredible variety of plants and animals, many of which are not yet known to science. These highly diverse pools of species constitute genetic reservoirs which have the potential for new sources of food, fiber, and medicinal and industrial products. Tropical forests are also a home for tribes who have an inherent right to their traditional lands, and whose knowledge of how to survive in the forest can provide a valuable lesson to cultures dependent on fossil fuels.

There are many causes of deforestation. Some of the reasons are political, social, and economic. Deforestation also results from improper management of cropland, pasture, or plantations that have replaced the forests. Improper management results in a rapid decline of productive capacity, which in turn necessitates additional forest clearing. Perhaps the most important cause of mismanagement of tropical lands is the lack of understanding of the factors which sustain production in tropical forest regions. Too often it has been assumed that techniques developed in and for temperate ecosystems are suitable for the wet tropics. This assumption has frequently led to disastrous consequences, especially when deforestation has been on a large scale.

The objective of this text is to develop an understanding of how and why tropical forests are different from temperate zone forests, of the factors important in sustaining productivity of tropical forests, pastures, cropland and plantations, and of how this knowledge can contribute to better management of tropical ecosystems. My hope is that better management of tropical ecosystems already influenced by man, will result in less need for disturbance of those forests that remain relatively untouched.

Carl F. Jordan
February, 1985

Acknowledgements

Chapter V of this book is an outgrowth of the 'Tropical Ecology Seminar Series', held at the Institute of Ecology, University of Georgia, between 1980 and 1982. Chapter III resulted from the 'Ecosystem Comparison Seminars' held in 1984.

A portion of the data for the La Selva, Costa Rica, site in Table III.3 was collected by Drs J. Luvall and G. Parker, while supported by a grant from the Ecosystem Studies Section of the National Science Foundation to the Organization for Tropical Studies.

I thank Drs B. Haines, F. Montagnini, and T. Whitmore for their valuable comments on the first draft manuscript, and K. Clark for final editing.

Introduction

A. Importance of nutrient cycling

Productivity,* the rate at which biomass is synthesized, is an important ecological parameter. Ecosystem productivity is an index which integrates the cumulative effects of the many processes and interactions which are proceeding simultaneously within the ecosystem. If productivity in a natural ecosystem changes little over a long period of time, it suggests that either the environment is unchanging, or that organisms or populations are compensating for changes which are occurring. If productivity changes dramatically, it could mean that an important environmental change is occurring, or that there has been an important change in the interactions of organisms within the ecosystem.

Man is often interested in maximizing productivity for commercial or other purposes. In agricultural ecosystems, management objectives may be to maximize productivity of food crops. In managed forests, the objective may be to maximize wood productivity. Often, ecosystems are managed for multiple objectives, such as production of both wood and wildlife. Sustained, but not necessarily high productivity has non-economic values, such as the prevention of soil erosion, the provision of habitats for valued species, and maintenance of an unspoiled ecosystem where people can enjoy wilderness experiences.

There are many factors that control productivity. Predation, disease, and competition are factors which often are important in regulating productivity of a species, a population, or a community of plants or animals. For example, bark beetles can reduce productivity of a pine forest, but woodpeckers can reduce the productivity of the beetles. In grain fields throughout the world, productivity often is limited by fungi which attack the grain and by weeds which compete with the crop plants.

Factors which are important in controlling productivity, and which have patterns caused by global and regional trends, are energy, water, and nutrients. Solar energy is the limiting factor in ecosystems with a high frequency of cloud cover. Solar energy is also indirectly limiting at high latitudes where short growing seasons limit the time when photosynthesis can occur. In deserts, of course, water is the limiting factor.

* 'Production' generally is used to refer to the process, and 'productivity' to the rate (Lieth and Whittaker, 1975).

1

It has been believed for many years that nutrients may often be limiting in the humid tropics in areas where forests are the natural vegetation (Richards, 1952). However, the evidence for the importance of nutrients and nutrient cycling in forests and cropland of the humid tropics has been scattered in the literature, and a strong case for the critical nature of nutrients in these ecosystems has never been made (Proctor, 1983). In recent years, a number of studies have been published which bear directly on the question. The objectives of this book are to review the studies pertaining to nutrient cycling in the humid tropics, to build a case for the critical nature of nutrients in this region, and to discuss the implications of nutrient scarcity for sustained productivity in both disturbed and undisturbed ecosystems.

Chapter I discusses the reasons why nutrients may be expected to be more critical in the tropics than at higher latitudes and why, within the tropics, nutrients may be more critical in lowland evergreen rain forest ecosystems than in drier and cooler environments. Chapter II shows how naturally occurring species in the wet tropics appear to have adapted to nutrient scarcity. Chapter III discusses how adaptations differ along gradients of temperature, moisture, and soil fertility. Chapter IV reviews the characteristics of the major nutrient element cycles, with a special emphasis on differences due to a tropical environment. Chapter V discusses how disturbances such as tropical deforestation for agriculture destroy nutrient-conserving mechanisms and change nutrient cycles, and how ecosystems recover after disturbance. The final chapter considers the implications of nutrient scarcity for management of ecosystems in the humid tropics.

B. Tropical deforestation

Tropical forests are being cut down at an increasing rate. There is disagreement as to the exact rates at which deforestation is occurring, partly because of the difficulty of defining what constitutes 'deforestation' (Sommer, 1976; Myers, 1980; Lanly, 1982; Lugo and Brown, 1982; Sedjo and Clawson, 1983). However, there is little disagreement that, whatever the rate, much of the world's primary tropical rain forest will have disappeared by the end of the 20th century if present trends continue. The main reasons for the cutting of tropical forests are: (1) the expanding populations of peasant farmers, (2) the need of many tropical countries for the capital gained from the export of timber and of agricultural products grown in previously forested areas, and (3) logging activities of multinational corporations (Myers, 1979, 1980).

What will happen as a result of increasing tropical deforestation? Will there be a drastic change as implied in the titles, *Amazon Jungle: Green Hell to Red Desert?* (Goodland and Irwin, 1975) and 'The Amazon Basin, another Sahel?' (Friedman, 1977), or will tropical rain forests be able to recover? Evidence developed in this book suggests that the answer depends, in part, on the availability of nutrients and the patterns of nutrient cycling in the wet tropics.

C. Meaning of 'nutrient cycling'

The term 'nutrient cycling' is used here instead of the frequently used term 'mineral cycling' because minerals do not really cycle. Although clay and rocks in the soil or on lake and ocean bottoms are composed of minerals, only those chemical elements such as calcium and phosphorus which are released when minerals are weathered actually cycle through an ecosystem. 'Biogeochemical cycling' is another term used to describe the cycling of elements, but it generally refers to cycles on a global scale and over geologic time, whereas the cycles discussed here are on the ecosystem scale and on the time scale of the lifespan of larger organisms. Element cycling would be an appropriate term if we were concerned with all the elements of the periodic table. However, here the interest is primarily in the cycles of major nutrient elements which are essential for life, such as nitrogen, phosphorus, potassium, and calcium. Therefore, the term nutrient cycling is used.

D. Data selection

The objective of this book is to discuss principles or concepts of nutrient cycling, particularly in relation to differences between the continually wet tropics and regions which have dry seasons, cold seasons, or both. Principles and concepts are generalizations formulated from comparisons of a large number of data points. The data points come from individual studies cited in the text or in other referenced publications.

Many of the studies cited are of ecosystem processes. In most cases the processes vary widely as a function of time and space. It is usually impossible to obtain a high level of confidence for an average value of an ecosystem process, because time and money limit the number of samples that can be taken. The problem is particularly acute in the tropics, where logisitics are often very difficult.

The large error terms in many ecosystem studies present a problem, in a review such as this one. Which studies should be considered adequate and which should be rejected as inadequate? Studies which cover a considerable length of time, perhaps several years or more, often are restricted in space. Studies which are extensive in space usually represent only one point in time. Limitations in methodology also are important, particularly when dealing with tropical studies, where common conveniences such as electricity, transportation, and dry shelter are often lacking.

There are almost no ecosystem process-level studies which are comprehensive in time and space. How, then, is it possible to find studies which are suitable for determining global trends in nutrient cycling? For the purposes of this book, no *a priori* judgements have been made on the adequacy of any published data. The only criteron used in selecting data is that it be published in a generally accepted source, such as a refereed journal or a PhD dissertation. The adequacy of each data point or set of data becomes apparent when it is plotted and compared with other data. Data points which are outliers from the trends suggested by the majority of data points can be suspect. These outliers may result from methodological problems, or there may be another reason to explain the deviation from the trend, for example, an ecosystem located on an unusual soil type.

Occasionally, there are generic difficulties with data when it is known that observations in a large number of studies are biased due to a common methodological problem. These types of biases, when they are known, are pointed out in the text. Otherwise, problems with individual studies are not discussed, unless it is clear that the data affect the pattern in an important way.

Chapter I

Factors which control nutrient cycles

This chapter discusses the reasons why nutrients can be expected to be more critical in the tropics, as compared to ecosystems at higher latitudes, and why, within the tropics, nutrients would be more critical in the hot, humid lowlands. Nutrient scarcity is caused, in part, by abiotic factors which follow global and regional patterns and, in part, by biotic factors which are strongly dependent on the abiotic factors.

A. Temperature

An important factor responsible for differences in nutrient cycling between tropical forests and forests at other latitudes is temperature. However, it is not extremely high temperatures which characterize the tropics. Temperatures in the lowland wet tropics are often lower than summertime temperatures in continental temperate regions. Instead, it is the distribution of temperatures throughout the year which is important (Walter, 1971). In the lowland tropics, high temperatures occur year-round, and biological processes dependent on high temperatures can go on continuously if moisture is not limiting.

High year-round temperatures have important effects on ecosystem process rates when other factors are not limiting. Constant high temperatures can mean that the growing season for plants never ceases. A continual growing season results in high annual rates of production, as will be shown later, and consequently in high annual rates of nutrient uptake by plants. Continuous production of leaves also results in continuous food availability for herbivores, and secondary productivity can be high. Consequently, in tropical rain forests there is potential for high annual rates of nutrient movement through food chains. High annual productivity also means high annual rates of nutrient return to the soil through leaf fall and through the death and defecation of herbivores and predators.

Nutrients returned to the soil in litter are solubilized through the action of decomposers. Because of year-round warm temperatures, tropical humid lowlands have the potential for continuous decomposer activity. This results in

5

continuous release of nutrients and a resultant high potential for leaching and recycling. Other microbiological processes important in nutrient cycling, such as nitrification, also are temperature dependent and can occur year-round in the wet tropics.

Because of year-round high temperatures, nutrient cycling processes can take place throughout the year in the lowland humid tropics, and annual nutrient cycling rates can be higher than those in regions where cold or drought interrupts these processes. These latter regions can occur within the tropical zone as well as at higher latitudes. Large areas of the tropics experience seasonal droughts of several months or longer and, although the tropics are often thought of as being continuously hot, there are high altitude regions within the tropical latitudes where temperatures are sometimes quite low. On the tops of the highest mountains there is often continual snow.

B. Moisture and temperature interactions

Year-round high temperatures will not result in continuous production and decomposition if seasonal droughts occur. The extent to which these processes slow down depends on the length and severity of the dry season. In many tropical regions there is a distinct dry season, while in others, such as the northwestern part of the Amazon Basin and parts of Malaysia, the 'dry season' may be just a few weeks when average weekly rainfall decreases slightly. In the least seasonal areas of evergreen wet forest, production occurs throughout the year. Where the length of the dry season is longer and little or no rain falls during the dry period, deciduous forests and savanna vegetation occur. During the dry season little growth takes place. Decomposition also slows, but fire can take the place of decomposer organisms in releasing nutrients from litter on the soil surface.

In the temperate regions, there is often seasonality of both temperature and rainfall. In some temperate regions, such as the eastern United States and northwestern Europe, rains are often frequent during periods of high temperature. However, because high temperatures and high rainfall coincide for only a few months each year, annual rates of production and decomposition are not as high as in the wet tropics. In other regions which have a 'Mediterranean' climate of hot dry summers and wet winters, nutrient cycles have different characteristics. Production and decomposition rates are slow, because temperature and moisture are not optimal at the same time. When temperatures are favourable for growth and decomposition, moisture is not available.

C. Biotic factors

Thus far in this chapter, the reason for potentially high annual rates of production, decomposition, and herbivory in the wet tropics have been given. In this section, process rates in the wet tropics are compared to those in other regions, to determine the extent to which potentially high rates in the wet tropics are realized.

1. Primary productivity

'Net production' by an individual plant is the amount of organic matter that it synthesizes and accumulates in tissues per unit time (Whittaker and Marks, 1975). Components of net production include leaves, wood, roots, fruits, flowers, bud scales, and organic substances leached or exuded from the plant.

The sum of the net production of all individual plants in a unit area of the Earth's surface is net primary productivity (Whittaker and Marks, 1975). In comparisons of net primary productivity, the time dimension must be carefully specified. Some studies report daily or monthly rates of net primary production during the growing season. Other studies give annual rates. Annual rates usually are less than 12 times the monthly growing season rates in regions where there is seasonal dormancy due to drought or cold.

There is no evidence that, nor any theoretical reason why rates of net primary production in the tropics should be higher than in other regions, on a daily or monthly basis during the growing season. Whitmore (1975) commented on the popular misconception that net assimilation rates are high in the tropics as follows:

> We can note that the luxuriousness and appearance of unbridled growth given by the vegetation of the perhumid tropics does not therefore arise from an intrinsically higher growth rate than exists amongst temperate species. The unfamiliar life-forms of palm, pandan, the giant monocotyledonous herbs of the Scitamineae (gingers, bananas, and their allies), and the abundant climbers make a vivid impression of 'vegetative frenzy' on the botanist brought up in a temperate climate. The appearance of rapid growth of pioneer trees of forest fringes and clearings, which forms the other part of the impression, does not result, as far as we yet know, from a particularly efficient dry-weight production or energy conversion but arises from the architecture of the tree which results from the capacity for unrestricted elongation of internodes and production of leaves in the continually favourable climate.

Although rates of primary production in the humid tropics do not appear greater than those at higher latitudes on a daily basis during the growing season, comparisons of net primary productivity on an annual basis can show a different trend. Since the humid tropical growing season is relatively long, there is the potential for greater annual production, simply because there are more days during which photosynthesis can occur.

Table I.1 shows net annual primary productivity for the major terrestrial ecosystem types of the world. The data clearly show that, on the average, tropical rain forests are the most productive terrestrial ecosystems in the world. However, there are some temperate forests and even some boreal forests which have higher rates of net primary production than some tropical forests (column headed 'Normal range', Table I.1). This is due to a lack of fine resolution in most classification schemes such that in as Table I.1. For example, the classification 'tropical rain forest' often includes dwarf heath forests and montane forests, as well as peat and other swamp forests, all of which are less

Table I.1
Net primary productivity in ecosystems of the world*

Normal range	Net primary productivity (dry matter)	
	Mean (of all studies) $g\ m^{-2}\ a^{-1})$	$g\ m^{-2}\ a^{-1})$
Tropical rain forest	1000–3500	2200
Tropical seasonal forest	1000–2500	1600
Temperate forest		
Evergreen	600–2500	1300
Deciduous	600–2500	1200
Boreal forest	400–2000	800
Woodland and shrubland	250–1200	700
Savanna	200–2000	900
Temperate grassland	200–1500	600
Tundra and alpine	10–400	140
Desert and semidesert scrub	10–250	90
Extreme desert – rock, sand, ice	0–10	3

* Reproduced by permission of Springer-Verlag from R.H. Whittaker and C.E. Likens, 1975, *The biosphere and man* (Ecological Studies, Vol. 14), pp. 305–328.

productive than the high forest on well drained intermediate sites (Brown and Lugo, 1984).

The values in Table I.1 may be underestimates of the real net primary productivity of ecosystems because some of the components of production are extremely difficult to measure. For example, in some ecosystems there may be a shedding of fine roots at the end of a growing season (Lyr and Hoffman, 1967). Data on fine root production and loss usually are not available (Hermann, 1977). Production of mycorrhizal fungi, which supply nutrients to roots and receive carbohydrates from them, is another component which can be significant (Odum and Biever, 1984) but is rarely measured.

On the other hand, the values in Table I.1 may be overestimates due to bias in the selection of forests that are studied toward those relatively large in biomass and high in productivity (Brown and Lugo, 1984). Comparisons such as those in Table I.1 also could be confounded by the inclusion of data from both successional and mature forests because successional forests are sometimes believed to have greater rates of growth than mature forests (Whitmore, 1975). Certainly growth rates of individual trees slow down as the trees mature, but mature trees die and are replaced by younger, faster growing trees. The idea that young forests are more productive than mature forests may arise in part from studies of even-aged plantations, where canopy closure does indeed result in a slowing down of the productivity of the entire forest. However, the heterogeneity of most naturally occurring forests may prevent such stagnation. Another reason for the belief that mature primary forests are less productive than regenerating forests of secondary species is that growth measurements are often in terms of volume or basal area instead of biomass (Jordan and Farnworth, 1980). The wood density of primary forest species may be two or three times greater than that of secondary species and may thereby compensate for a smaller volume increment.

Care must be taken not to confuse net annual increment of a forest with net primary productivity. A successional forest can show a large annual increment in biomass and a mature forest no annual biomass increment, yet both can have the same net primary productivity. The reason is that a large part of the production of the successional forest is accumulated as standing stock of biomass, whereas much of the production of the mature forest passes quickly to the decomposers.

Net primary productivity in Table I.1 is the natural productivity of eco-systems without subsidies such as fertilization or irrigation. Short-term produc-tivity can be higher than the values in Table I.1 when nutrients or water are added. Since such subsidies as water and fertilizer are used irregularly on a global scale, ecologists prefer to use net primary productivity of unmanaged ecosystems to detect global patterns of productivity and the factors which affect it.

Although total net primary productivity is often higher in tropical forests than in temperate or boreal forests, these higher rates seem to be due almost entirely to greater leaf production. Wood production does not differ signifi-cantly along a gradient from high latitudes to the tropics in mesic lowland or lower montane late successional or mature hardwood forests (Fig. I.1; Jordan,

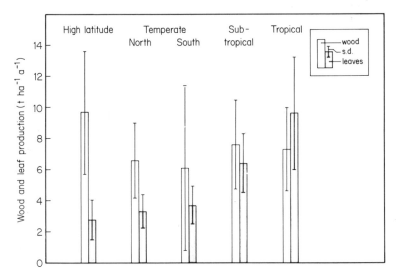

Fig. I.1 Averages and standard deviations of wood and leaf production for mesic hardwood forest ecosystems at various latitudes. Averages are based on 60 studies of wood production and 135 studies of leaf litter production. Individual data values and references are listed in Jordan (1983).

1983). It appears that in the tropics relatively more of the photosynthate produced by trees is allocated to leaves, whereas at higher latitudes, relatively more is allocated to wood. The reason for this pattern of productivity is not clear, but because plants with large structure can intercept more light, high wood production relative to leaf production could be of adaptive advantage in regions where light is or has been limiting. The annual amount of available light may have been critical during the late Tertiary period at high latitudes due to the shortening of growing seasons (Jordan and Murphy, 1978) which resulted from the world-wide lowering of temperature during this period, evidence for which is summarized by Wolfe (1978).

2. Decomposition

Leaf and wood litter reaching the forest floor decay and gradually become incorporated into the upper horizons of the mineral soil through the activity of soil organisms. The organic matter on the soil surface and mixed in with the mineral soil is the major storage reservoir for soil nitrogen and sulfur, and it is also an important reservoir for phosphorus, calcium, potassium, and magnesium, and other nutrients (see Chap. IV). Release of nutrients from the decaying organic matter in the soil is a critical step in ecosystem function. If nutrients are released too fast, they can be lost through soil leaching or volatilization. In contrast, if decomposition is too slow, insufficient nutrients are made available to plants, with the result that tree growth can be inhibited.

(a) Rates of decomposition

There are several general approaches to measuring rates of organic matter decomposition. One involves determining the respiration rate through measurements of the rate at which carbon dioxide evolves from decomposing material (Medina and Zelwer, 1972). A major problem with this method is the difficulty of separating microbial decomposition from root respiration (Singh and Gupta, 1977; Medina et al., 1980). Another approach is simply measuring weight loss of decomposing material. There are numerous problems with this approach also (Swift et al., 1979), but there have been many studies of this type and comparisons can yield some insights.

Results of litter weight loss studies are usually reported in terms of the exponential decomposition constant, k. This term is related to the disappearance of organic material by the equation

$$\frac{X}{X_o} = e^{-kt}, \tag{1}$$

where X_0 is the original amount of litter, X is the amount of litter remaining at a later time, e is the base of natural logarithms, and t is the time elapsed between X_0 and X. (Derivation given by Olson, 1963.) 'Litter' can consist of leaf litter only, or it can include woody material as well, and it is necessary to specify the fraction measured since woody litter decomposes much more slowly (Lang and Knight, 1979).

The time required for a certain proportion of the litter to decompose can be calculated from equation (1). For example, if it is desired to know the length of time for half the original stock of litter to decompose, X is taken to be half of X_0, and the equation is solved for t.

In an ecosystem where rates of litter fall and decomposition are approximately equal, and the stock of litter is thus in approximate steady state, k is related to standing stock of litter (X) and rate of litter fall (L) by the equation

$$k = \frac{L}{X} \quad \text{(Olson, 1963)}. \quad (2)$$

In steady state, litter fall is equal to decomposition loss, and the stock of litter is related to decomposition by k.

A major shortcoming of using k values to express rates of decomposition is that it suggests that the rate at which a particular piece of litter decomposes is constant from the time the litter lands on the ground to the time that the last organic molecule is respired. Actually, decomposition rates can change significantly depending on the season of the year, the chemical nature of the remaining undecomposed material, and other factors (Swift et al., 1979). Nevertheless, k values have often been used to compare rates of decomposition.

The comparisons of k on a global scale by Olson (1963), as well as the field studies by Jenny et al. (1949) and Madge (1965), suggested that decomposition rates in the tropics are higher than those at other latitudes. This idea has been widely held for several decades, but Anderson et al. (1983) have recently challenged the conclusion that the decomposition rate of litter is higher in the tropics. They present a series of k values for leaf litter from a variety of habitats and conclude that variation within single regions is too great to permit statements about differences between tropical and temperate latitudes.

While it is true that their range of k values within a region is too great to show statistically significant differences between regions, nevertheless, there is still reason to believe that decomposition rates are higher in the wet tropics. Let us first consider rates of litter fall from Jordan (1983) which are summarized in Fig. I.1. While there are problems with litter fall measurements, there are fewer problems than with decomposition measurements. If it is accepted that leaf litter fall rates in the wet tropics are higher than rates in other regions, then decomposition rates, in terms of grams of carbon released per unit area per unit time, must be higher in the tropics, otherwise there would be an infinite build-up of the litter layer.

The amount of litter, X, can complicate comparisons of k. When decomposition values are expressed in terms of k, similar values in different regions could occur despite differences in rates of litter fall and decomposition. For example, if litterfall (L, equation 2) in the tropics were double that in the temperate zone, but standing stock of litter (X) also were doubled, the decomposition constant k would be the same in both regions (equation 2), even though absolute rates of decomposition were higher in the tropics. However, in general, stocks of litter probably are not higher in the tropics. The extensive data compiled by Zinke *et al.* (1984) on soil organic carbon (Fig. I.4) suggest that, on the average, stocks of soil organic matter are lower in the tropics than at higher latitudes, under comparable moisture regimes. Thus, measurements of soil organic matter stocks and leaf litter fall, which are easier to measure than litter disappearance, both suggest that decomposition does follow the trend suggested by global patterns of temperature and moisture; that is, that annual rates are highest in the continually humid tropics.

One of the reasons that many tropical decomposition studies show relatively low values of k is that the leaves of tropical trees are often quite sclerophyllous (Mooney *et al.*, 1984). Sclerophyllous leaves, i.e. leaves that are tough and have a low ratio of area to weight, are relatively resistant to decay, at least during the first stages of decomposition. Jenny *et al.* (1949), who concluded that decomposition is higher in the tropics, used the same types of leaves in both tropical and temperate study sites, thereby eliminating the effect of substrate type on measurements of decomposition rate.

While the evidence for higher annual rates of decomposition in the tropics may remain inconclusive, the lack of strong statistical trends may relate more to the difficulties of direct measurement of decomposition than to lack of world patterns.

(b) Nutrient pathways

The pathways of nutrient flow through the soil system are complex, and nutrient loss or recycling depends to a large extent on the dynamics of nutrients along these pathways (Coleman, 1976; Coleman *et al.*, 1983). Nutrients released from decaying leaves and wood as well as from animal carcasses usually do not move directly to the soil or roots of trees, but instead pass through a whole series of small-scale cycles within the organic matter portion of the soil (Fig. I.2). These are similar to the 'spirals' that occur in stream ecosystems (Webster and Patten, 1979; Newbold *et al.*, 1982). In the case of decomposing leaves, the cycles often being with soil arthropods which chew the leaves. As particles pass through their digestive systems, the complex organic compounds are changed to simpler compounds which are more readily utilized by other soil organisms. Decomposition of wood and leaves also can begin with invasion of the tissue by bacteria and fungi. These microorganisms quickly immobilize any soluble cations in the tissue and alter the nature of the substrate, sometimes rendering it more resistant. Resistant substrates often favor fungi for continued colonization. As compounds excreted from fungal hyphae continue to break down complex organic compounds in the leaves and wood, the nutrient concentration in the litter is again increased and bacterial

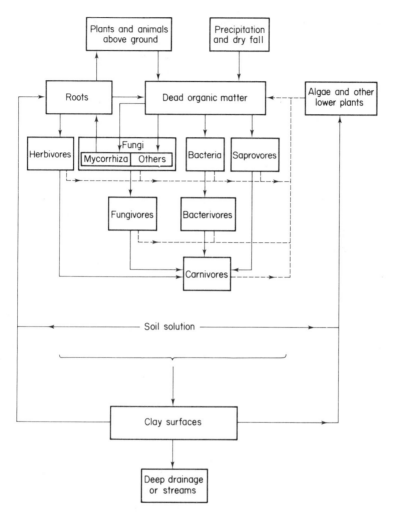

Fig. I.2 Diagram showing some of the possible flows of nutrients through the below-ground portion of an undisturbed forest. Dashed lines indicate return of organisms to dead organic matter pool. Soluble nutrients may be excreted by many of the organisms, or may be released during trophic transfers, Nutrients in soluble form may not follow trophic pathways. Soluble nutrients may be taken up by algae, or other plants, or they may be leached into the mineral soil where they can be absorbed on clay, taken up by roots, or leached through the subsoil or to drainage streams. Gaseous fluxes such as nitrogen fixation and denitrification are not indicated. Deforestation greatly simplifies the cycles, and increases the proportion of nutrients in soluble form. [Adapted from Persson *et al.* (1980) by permission of Ecological Bulletins.]

colonization is once again favoured. The succession of organisms during decomposition is very complex, the species involved numerous, and almost every case is different in details (Swift *et al.*, 1979).

The relative digestability of litter is influenced, in part, by such factors as carbon : nitrogen ratio, and lignin content (Swift *et al.*, 1979). A low C:N ratio (less than about 25:1) usually indicates a high concentration of all nutrients and, consequently, a favorable environment for high bacterial activity. A low carbon : nitrogen ratio and a low lignin content also make the litter a more attractive food source to a variety of scavengers such as millipedes, wood lice, fly larvae, springtails, and various types of worms. As these animals pass the debris through their digestive systems, the litter is broken into smaller pieces and the shredding process allows another group of fungi and bacteria to invade the litter. These organisms, in turn, continue to make the litter more palatable for a number of other decomposers. Bacteria and fungi themselves may be ingested by larger organisms.

Nutrients from living vegetation also enter the decomposer cycle. Roots of living trees are fed upon by nematodes and cicada larvae living in the soil. In some ecosystems, a relatively large amount of fine root hairs are shed yearly, and these may pass through the decomposer cycle in the same manner as litter above ground (Harris *et al.*, 1977).

Nutrient flow usually follows flow of energy or carbon through the decomposer cycle. Exceptions are the mycorrhizal fungi which form symbiotic relationships with roots. Mycorrhizae obtain most of their carbohydrates from the roots and, in turn, they increase the capacity of roots for nutrient uptake (Ruehle and Marx, 1979).

Figure I.2 is a generalized scheme of nutrient flow through the below-ground community. Structure and function of the community, and the species comprising the community, will be different in almost every ecosystem. For example, earthworms and their intestinal microflora which appear to play an important role in the digestion of organic matter form an association which differs in major characteristics in different global regions. The differences are due largely to the nature of the soil organic matter (Lavelle, 1983). In cold regions, litter decomposes slowly and constitutes an abundant and reliable food source of high energy content for decomposers. The litter contains hydrosoluble organic compounds and energy substrates that can be transformed into simple compounds by a low level of microbial activity. In contrast, soil humic reserves, though abundant, are rarely accessible, and relatively little soil is ingested. Microbial activity appears to be insufficient to produce enough primary substances from stable humus to sustain earthworm nutrition. The earthworm communities are, consequently, composed predominantly of detritivorous species. In tropical areas the situation is very different. Rapidly decomposed litter does not provide an abundant and reliable food source, and is soon deprived of its hydrosoluble energetic content by heavy, warm rains. Soil humic reserves, however, constitute an abundant food source. Lavelle (1983) has shown that, as a consequence, tropical earthworm communities are frequently dominated by geophagous species.

Another important difference between tropical and high latitude pathways of decomposition is the importance of termites in the tropics (Matsumoto, 1976, 1978). Nutrient recycling in tropical ecosystems probably is affected by termites which attack senescent trees. Some species of termite carry material back to their nests, where it is decomposed by a fungus (Wood, 1978). The fungus is used as a food source for the termite colony. 'Leaf cutter' ants also cultivate similar 'fungal gardens' on leaf fragments cut from living forest trees (Stradling, 1978).

3. Balance between productivity and rates of decomposition

The amount of non-living organic material in the soil is a function of the balance between productivity and the rate of decomposition. To illustrate, we can consider an experiment in which organic matter is removed from an area of forest floor (time 0, Fig. I.3). Because there is no soil organic matter, there are very few decomposers. As leaves and other organic matter fall onto the experimental plot there is a net accumulation of organic matter because the few decomposers present are not able to break down the organic matter as fast as it comes into the plot. However, as the organic matter, which is the food supply for decomposers, builds up, the populations of decomposers also increase until the rate of decomposition just equals the rate of production. In Fig. I.3, this occurs at a soil organic matter biomass of 200 g/m². In many forests, leaf fall occurs in pulses rather than uniformly throughout the year. The effect of such input pulses is shown as a serrate curve in Fig. I.3.

Zinke *et al.*, (1984) and Post *et al.*, (1982) have plotted soil organic carbon on a global basis as a function of climate, using 3500 soil profiles from natural or little disturbed sites (Fig. I.4). The pattern is superimposed on the life-zone scheme of Holdridge (1967). In this scheme the mean annual biotemperature is plotted on the vertical axis on the left side of the figure. Biotemperature is the annual mean temperature, with 0 substituted for any temperatures below 0°C. The left diagonal axis of the triangle is the ratio between mean annual potential

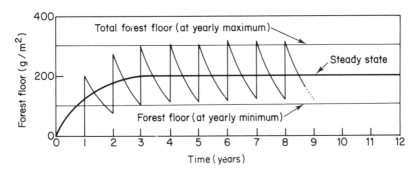

Fig. I.3 Stock of soil organic matter under conditions of steady input and loss of litter compared with step-wise curve for additions and losses of litter in an idealized deciduous ecosystem. The steady state stock is proportional to the decomposition rate (see equation 2). [Adapted from 'Energy storage and the balance of producers and decomposers in ecological systems' by J.S. Olson *Ecology*, 1963, **44**, 322–331, Copyright © 1963 by Ecological Society of America. Reprinted by permission.]

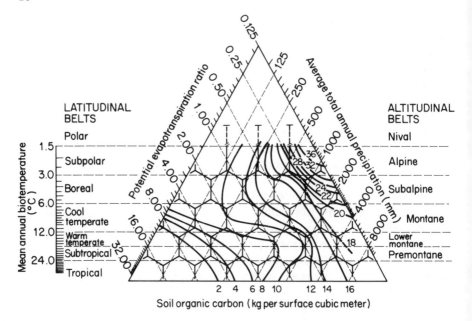

Fig. I.4 Contours of soil carbon accumulation plotted on world life-zone classification (from Zinke *et al.*, 1984, and Holdridge, 1967). The formations for each hexagon are as follows, from left to right in each latitudinal region and altitudinal belt: *subpolar—alpine*: dry tundra, moist tundra, wet tundra, rain tundra; *boreal—subalpine*: desert, dry scrub, moist forest (Puna), wet forest (Paramo), rain forest (Rain Paramo); *Cool temperate—montane*: desert, desert scrub, steppe, moist forest, wet forest, rain forest; *warm temperate and subtropical—lower montane and premontane*: desert, desert scrub, thorn woodland, dry forest, moist forest, wet forest, rain forest; *tropical*: desert, desert scrub, thorn woodland, very dry forest, dry forest, moist forest, wet forest, rain forest. [Redrawn from Zinke *et al.*, 1984.]

evapotranspiration and mean annual precipitation. The right diagonal axis is the average total annual precipitation. The horizontal axis is the stock of soil carbon in kilograms per square meter. Soil carbon is roughly 50 percent of soil organic matter.

In Holdridge's original figure, vegetation types are indicated within the hexagons inside the triangle. They range from hot desert in the lower left-hand corner and dry tundra in the uppermost hexagon on the left, to rain forest in the lower right corner and rain tundra in the uppermost hexagon on the right. The vegetation types for each hexagon are given in the legend of Fig. I.4.

There are several problems with the classification scheme of Holdridge (1967), one of which is that the scheme takes into account variations in temperature amd moisture, but not soil nutrients. Nevertheless, all classification schemes are imperfect in one way or another, and the life-zone scheme is one of the most convenient for abstracting world patterns.

An important trend illustrated by Fig. I.4 is the effect of a pronounced dry season in the subtropics and tropics on the balance between productivity and decomposition rate. The effect is shown most strongly by the 10 kg/m^2 carbon isoline. Moving downward from the polar belt, the line shifts sharply to the right near the transition between the cool temperate and warm temperate belts. The vegetation types represented by the hexagons just below this shift are dry and moist warm temperate and subtropical forests. The abrupt shift to the right of the isoline may be explained as follows: a dry season in the warmer zones has a greater effect on annual rates of production than on annual rates of decomposition. During dry seasons, the production of leaves ceases almost entirely, while decomposers in the soil are less affected and are able to maintain at least some activity. Consequently, the standing stock of carbon in the soil of subtropical seasonally dry forests is lower than the stock in temperate moist forests, although all may have similar annual rates of production.

The most important trend illustrated in Fig. I.4, in relation to nutrient cycling in the tropics is the increase in stocks of soil organic matter as one moves up along the right-hand diagonal, from rain forest at the bottom to rain tundra at the top. The lower stocks in the wet tropics occur despite the fact that annual productivity is higher in the wet tropics. These data suggest that the decomposition rate changes more rapidly as a function of temperature than does productivity. The world patterns of soil organic matter shown in Fig. I.4 conform closely to trends predicted by early Russian workers (reviewed by Volobeuv, 1964) based on production and decomposition data gathered in the first decades of the 20th century.

The evidence that the balance between production and decomposition results in a relatively low standing stock of soil organic matter in the tropics, under conditions where moisture and nutrients are not critical, has important implications. It means that, when a tropical rain forest is cut and the litter input to the forest floor is eliminated, the rate at which soil organic matter disappears is much more rapid than it would be at higher latitudes. Because soil organic matter is such an important component in the storage and supply of nutrient elements, deforestation in the tropics can have a much greater effect on the supply of soil nutrients than deforestation at higher latitudes. This effect will be discussed more fully in later chapters.

4. Herbivory

Herbivory is another important biotic factor which influences nutrient cyles (Seastedt and Crossley, 1984). However, in contrast to production and decomposition, comparison of herbivory rates at different latitudes is difficult because total ecosystem herbivory is rarely studied and because large grazers are often studied under artificial conditions (Barnes, 1983).

Because of continual high temperatures and high humidities in the humid tropics, it might be anticipated that the activity of insect herbivores would be high. However, at least in naturally occurring forests, the little evidence available suggests that insects rarely defoliate large areas (Lugo et al., 1974), although defoliation of individual trees often occurs (Janzen, 1983).

Ecologists have speculated that there may be a relationship between the apparent resistance of tropical rain forests to large-scale defoliation and the high species diversity of these communities (Orians *et al.*, 1974; UNESCO, 1978). High species diversity means that distances are large between host plants or clumps of plants (Hubbell, 1979). When insects begin searching for additional food sources, it may be difficult to find another suitable host. The idea that large-scale outbreaks of herbivores are rare in tropical rain forests arises from the hypothesis that the high diversity of tropical trees is *caused* by herbivores. Evidence that herbivores may play a role in causing the high diversity of rain forest trees has been reviewed by Gilbert (1980) and Connell *et al.* (1984).

In contrast to the humid tropics, plant species diversity is much lower in drier tropical regions. In areas such as the African savannas, where a few species of grasses and herbs cover vast areas, large grazing herbivores are important regulators of ecosystem structure and function (Sinclair and Norton-Griffiths, 1979). For instance, they stimulate productivity of senescent grasslands by grazing (McNaughton, 1976).

In agricultural ecosystems of the humid tropics, where species diversity is not as high as in natural forests and where many crops lack the natural defenses of wild species, year-round warm, moist conditions are highly conducive to exponential growth of herbivore populations. Consequently, the potential for crop damage is much higher than in regions where a cold or dry season set back the populations of insects (Janzen, 1973).

Lack of species diversity and year-round growing season are reasons that the conversion of tropical rain forest to crop monoculture is potentially disastrous. For example, in tropical mahogany plantations the larvae of the mahogany shoot borer kill or distort the shape of young plants. It is almost impossible to establish large-scale mahogany plantations where shoot borers are present, although these insects do only minor damage in native forests (Hodges, 1981). Lack of species diversity also may contribute to destruction of tree plantations by weevils and miners in South-East Asia (Hodges, 1981) and by leaf cutter ants in South America (Fearnside and Rankin, 1980).

Although fungal, bacterial, and viral diseases affect host plants differently than herbivores, they respond in the same way to conditions of year-round high temperatures and high moisture. Spread of such diseases is well documented in tropical monocultures. For example, between 1939 and 1953, banana diseases almost completely destroyed plantations in eastern Panama and brought extreme poverty to the region (Gordon, 1982). Other diseases have been highly destructive in large-scale plantations of cacao, rubber, pulpwood, and coffee. Diseases are especially difficult to control when the host plants are propagated clonally, since this provides the parasite species with a genetically uniform food source to which it can more easily adapt. Genetic uniformity also may be the reason that conifer plantations in the tropics often appear less resistant to disease than naturally occuring stands (Brown, 1981). An extensive review of pests and diseases in tropical forests and plantations has been given by UNESCO (1978).

D. Influence of biotic factors on leaching and weathering

The evidence presented so far suggests that constant high temperatures and high humidities result in high rates of decomposition. This in turn results in a high annual rate of nutrient release which makes possible the high annual rates of primary production observed in the humid tropics. The environmental conditions which favor high annual rates of nutrient recycling and biomass production also create the potential for high rates of leaching and weathering of tropical soils.

1. Leaching

Some of the nutrients in the soil are stored on the surfaces of clays and colloidal organic matter (humus) (Brady, 1974). These surfaces are negatively charged. Positively charged ions, such as K^+, Ca^{++}, and NH_4^+, are attracted to the negative surface charges. The ability of a soil to retain cations on the surface of clay and humus particles is called cation exchange capacity. Soils with a high clay content or a high organic matter content often have a high cation exchange capacity, but the type of clay mineral present is also important. Highly weathered clay minerals such as kaolinite, whch is common in geologically old areas of the tropics such as the Amazon Basin, have relatively low cation exchange capacities (Sanchez, 1976).

Positively charged nutrient ions exchanged on the surface of clays can be replaced by hydrogen ions. Hydrogen ion availability in tropical soils is potentially high because carbonic acid is continuously produced through the respiration of soil microorganisms and roots. Decomposer respiration plus root respiration is called 'soil respiration'. When the soil is wet, carbon dioxide (CO_2) from soil respiration combines with water to form carbonic acid (H_2CO_3) which dissociates into bicarbonate (HCO_3^-) and a positively charged hydrogen ion (H^+):

$$CO_2 + H_2O \rightarrow H_2CO_3 \qquad\qquad (3)$$
$$H_2CO_3 \leftarrow H^+ + HCO_3^- \qquad \text{(Johnson } et\, al\, 1977)$$

The hydrogen ions can displace nutrient cations on the soil colloids. The bicarbonate reacts with cations released from the colloids, and the products leach down through the soil profile. Whether leached cations actually percolate out of the soil horizons depends on characteristics of the subsoil (Cole et al., 1975).

Although water, in a general sense, is the main agent of chemical weathering (Bowen, 1979), it is the acidity of the water which governs the rates of leaching and weathering (Johnson et al., 1975, 1982b).

Nitrogen transformations also are a source of hydrogen. During decomposition of organic matter, such as leaf litter on the forest floor, organically bound nitrogen is converted to ammonium (NH_4^+) which can be taken up by plants or converted to nitrate (NO_3^-) through the process of nitrification. When ammonium is taken up by plants a hydrogen ion must be released to maintain electrical neutrality (Nilsson *et al.*, 1982). Hydrogen is also released during the processes of nitrification:

$$2NH_4^+ + 3O_2 \rightarrow 4H^+ + 2H_2O + 2NO_2^- \tag{4}$$
$$2NO_2^- + O_2 + 2NO_3^-$$

(Delwiche 1977)

The hydrogen released by these processes can replace nutrient cations on soil exchange surfaces, and nitrate ions are available to react with the displaced cations.

Hydrogen released into the soil as a result of biological activities reacts with silicate clays releasing exchangeable aluminum (Sanchez, 1976). Because aluminum is prevalent in many tropical soils, aluminum, rather than hydrogen, is the dominant cation associated with soil acidity in these regions (Coleman and Thomas, 1967).

Other processes also can be a source of ions which contribute to nutrient leaching. Sulfate can be carried into ecosystems either in rainfall or in dry deposition (Johnson *et al.*, 1982a). Sulfate may be formed in the soil from gaseous sulfur compounds by microbiological activity (Ivanov, 1981). Organic acids, from decomposing leaf and wood litter, also play a role in soil acidity (Johnson *et al.*, 1982b). These acids can be important in nutrient leaching processes when they are carried down into the mineral soil by percolating water (Binkley *et al.*, 1982).

The relative importance of various anions as leaching agents may change as a function of latitude, altitude, or the nature of the underlying bedrock (Johnson *et al.*, 1975). Frequently, carbonic acid is important in tropical areas, because the year-round soil respiration results in high production of CO_2. The importance of carbonic acid in a wet tropical forest can be seen from the comparison of bicarbonate anions with other anions (Fig. I.5). Bicarbonate has a lower volume weighted concentration in non-tropical regions.

The sum of the anion concentrations at each site (Fig. I.5) do not show the much greater leaching potential in the tropical rain forests. To determine the total yearly cation leaching potential, the anion concentrations must be multiplied by the amount of water percolating through a unit area of forest floor per year. Although the annual percolation at each site is unknown, the average annual precipitation values at the rain forest, temperate, alpine, and taiga sites are 430, 136, 270, and 270cm, respectively (Johnson *et al.*, 1977). The relative amounts of precipitation at the four sites suggests that the quantity of percolating water is highest in the rain forest site, since evapotranspiration in the tropical rain forest may be relatively low due to high humidity. Thus, leaching potential in the rain forest, compared to other regions, is greater than suggested by the relative sizes of the curves in Fig. I.5.

The importance of nitrate as a potential leaching agent appears relatively low in Fig. I.5. This may be due to inhibition of nitrification in undisturbed forests (Jordan *et al.*, 1979a). However, when forests are cut and the inhibition mechanisms destroyed, nitrate becomes relatively important in leaching pro-

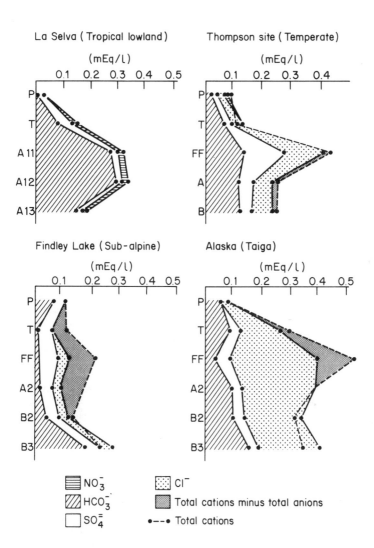

La Selva (Tropical lowland)

Thompson site (Temperate)

Findley Lake (Sub-alpine)

Alaska (Taiga)

NO_3^-

HCO_3^-

$SO_4^=$

Cl^-

Total cations minus total anions

●--● Total cations

Fig. I.5 Volume weighted average anion and total cation concentrations in soil solutions from four contrasting ecosystems. The La Selva site is a tropical rain forest in lowland Costa Rica. The Thompson site has a temperate climate and is in the foothills of Washington state at an elevation of 210 m. The Findley Lake site is at an elevation of 1110 m in the mountains of Washington and represents a subalpine environment. The Alaska site is at Petersburg at an elevation of 150 m and represents a taiga ecosystem. P, precipitation; T, throughfall, FF, forest floor; A and B are soil horizons. [Adapted from Johnson *et al.* (1977) with permission of the Regents of the University of Colorado from *Arctic and Alpine Research.*]

cesses, and may be responsible for the flush of cations often observed after forest clear-cutting (Vitousek *et al.*, 1979, Sollins and McCorison, 1981).

Organic acids are less important in the leaching processes of tropical forests than in those of high latitude forests (Johnson *et al.*, 1982b) probably because they decompose rapidly in warm, wet environments.

The evidence presented here suggests that continual hydrogen ion availability in the soils of wet tropical regions results in high leaching potential for nutrient cations. High hydrogen availability also allows soluble phosphorous to react with iron, aluminum, and manganese to form insoluble compounds (Sanchez, 1976). These compounds are not readily available to many crop plants and, consequently, phosphorous deficiency is common in tropical agricultural systems (Olson and Engelstad, 1972).

The point emphasized in this section has been that, because of the year-round warm wet conditions in tropical rain forests, processes that have the *potential* to contribute to nutrient cation leaching and phosphorous immobilization go on almost continuously. Whether nutrient loss and immobilization actually *are* important in undisturbed rain forests depends, to a large extent, on the effectiveness of the nutrient conserving mechanisms discussed in the next chapter.

2. Weathering

Constant warmth and moisture also produce potentially high annual rates of weathering. Weathering occurs when hydrogen in the soil solution reacts with minerals in the soil or bedrock, resulting in removal of nutrient elements. For example, feldspar is an aluminosilicate (aluminum and silica compound) containing nutrients such as sodium, potassium, and calcium. When it is hydrolyzed, the nutrients are removed from the aluminosilicate. The soluble nutrients can be adsorbed by soil colloids, used by plants, or removed in the drainage water. Since the reaction rates are temperature dependent, the weathering processes are more important in lowland tropical areas than at higher latitudes or altitudes.

A general sequence of mineral transformation and nutrient element loss during weathering is shown in Fig. I.6. As severity and length of weathering time increase, the proportion of nutrient elements in the minerals decreases. The most highly weathered clay minerals, such as kaolinite, consist only of silica, aluminum, hydrogen, and oxygen. Continued weathering and leaching of kaolinite results in the removal of silica, leaving oxides or iron and aluminum in the upper soil horizons. This can result in the formation of a hard-pan or impermeable crust of 'laterite' when the iron and aluminum oxides are exposed to air and dry out (Jenny, 1980).

Observations on laterite formation date back to 1800. According to Jenny (1980),

> In the red earth country of India's Malabar Coast, F. Buchanan, M.D., in 1800, observed Indians cutting slabs out of soft red and ivory mottled clay strata, air-drying them to a hard rock, and using them as building stones. Buchanan called brick and hardened stone mantles laterite, from the Latin word 'later' meaning brick.

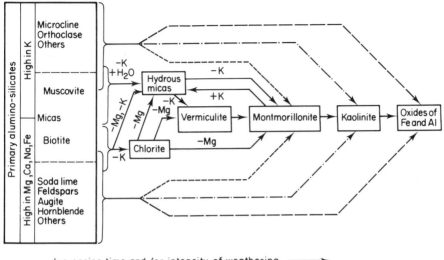

Increasing time and /or intensity of weathering ⟶

Fig. I.6 A generalized sequence of mineral transformations as a result of weathering. Unweathered minerals in rocks are shown on the left. Losses of nutrient elements during transformation to clay minerals and during weathering of clay minerals are shown. The ultimate products (oxides of iron and aluminum) on the far right, occur when all the nutrient cations and silica have been removed. – – – –, hot wet climates, silica, leaching important; –.–, rapid removal of bases; ......, slow removal of bases, magnesium present. [Adapted from Brady (1974) with permission from Macmillan Publishing Co.]

In recent years, the increasing rate of tropical deforestation has led to the fear that the exposure of soils to air after clearing will result in the formation of extensive hard crusts or 'red deserts' (McNeil, 1964). There is little evidence, however, that laterite formation is a major problem in the wet tropics (Sanchez, 1976). On the Jarí plantation in Pará, Brazil, in the eastern Amazon Basin, large areas of oxisol and ultisol have been exposed after clear cutting. Because of the pronounced dry season in the area, laterite might be expected to develop near the surface of these soils. However, after the primary forest is cut, the soil is covered with plantations or natural successional vegetation, which prevents the severe drying necessary for laterite formation (C.F. Jordan, personal observation). In fact, laterite is used for building roads and air-strips in the Amazon region, but, in order to maintain them free from invading vegetation, they must be occasionally bulldozed or treated with other heavy machinery.

Since high rates of weathering occur more frequently in tropical regions than at higher latitudes, there is a general world pattern of occurrence of the different types of clay minerals (Millot, 1979). At the highest latitudes, relatively unweathered clay minerals such as illite and chlorite often predominate. In temperate zones, vermiculites are common, and in Mediterranean and seasonal tropics, montmorillonite (smectite) frequently occurs. In the humid tropics, kaolinite and gibbsite, both highly weathered clay minerals, are frequently encountered.

Although the least weathered minerals often occur at the highest latitudes, while the highly weathered minerals occur in the wet tropics, there are exceptions. For example, the data of Melfi *et al.* (1983) do not support the idea that weathering rates are unusually high in the tropics. However, their study sites were located in either semi-arid savannas or temperate regions of Brazil, where drought or cold interrupts chemical weathering processes. Exceptions also occur in mountainous regions, where erosion exposes unweathered bedrock (Fyfe *et al.*, 1983). The tendency for highly weathered clay minerals to occur chiefly in the tropics is most pronounced in the continually, or almost continually hot, wet lowland forests, where there has been no uplift of bedrock over long periods of geologic time. In central Amazonia, where weathering processes have been altering the clay minerals for 100 million years or more (Kronberg *et al.*, 1982), the minerals are principally quartz, kaolinite, gibbsite, goethite, and hematite (Kronberg *et al.*, 1979). These forms are among the last in the mineral weathering sequence (Jackson *et al.*, 1948). This deep, intense, and long-term weathering also results in the formation of deep layers of partially decayed bedrock (saprolite; Jenny, 1980) which prevents root penetration to the unweathered bedrock below.

It is because of year-round leaching and weathering in the lowland humid tropics, that a very large proportion of the soils in these regions are nutrient-poor oxisols and ultisols. In the humid tropics in general, 63 percent of the soils fall in these categories. In humid tropical America, 82 percent of the soils fall into these categories. In the Amazon Basin, only 6 percent of the soils have no major nutrient limitations (Sanchez, 1981).

3. Comparisons of nutrient balance

A major point developed in this chapter is that due to the continuous biological activity, a potential for high rates of nutrient leaching and mineral weathering exists in the humid tropics. Although many tropical soils are highly weathered, it appears that the leaching potential often is not realized, at least in undisturbed ecosystems. Table I.2 gives rates of calcium runoff for 18 ecosystems throughout the world, listed in increasing order of calcium loss through runoff. Some of the ecosystems with the highest potential for nutrient loss (lowland rain forests; Table I.2, ecosystems 1–4) actually have the lowest rates of calcium loss. Other tropical rain forests (ecosystems 17 and 18) have the highest rates of calcium loss. The differences among the rain forests are due to the nature of the parent rock. Ecosystems 17 and 18 are on calcareous rock and the leaching losses are probably partially offset by rapid weathering and release of calcium from the subsoil.

Nutrient input from the atmosphere also compensates for ecosystem nutrient losses. Nutrients enter from the atmosphere dissolved or suspended in rainwater, or as aerosols which settle out of the atmosphere as 'dry fall'.

Table I.2

Runoff (R), atmospheric input (A), and difference (A−R) for calcium in various ecosystems arranged in increasing order of calcium runoff

Formation or association, and location	Runoff (R)	Input (A)	A−R	Author
	Calcium (kg ha⁻¹ a⁻¹)			
1. Tropical rain forest, Malaysia	2.1	14.0	+11.9	Kenworthy, 1971
2. Rain forest, Amazon Basin in spodosol	2.8	16.0	+13.2	Herrera, 1979
3. Evergreen forest, Ivory Coast	3.8	1.9	−1.9	Bernhard–Reversat, 1975
4. Rain forest, Amazon Basin oxisol	3.9	11.6	+7.7	Jordan, 1982a
5. Pine forest, North Carolina	4.1	6.5	+2.4	Swank and Douglass, 1977; Johnson and Swank, 1973
6. Douglas fir forest, Washington	4.5	2.8	−1.7	Cole et al., 1967
7. Coniferous forest, north Minnesota	4.5	3.1	−1.4	Wright, 1976
8. Mixed coniferous forest, New Mexico	4.9	7.6	+2.7	Gosz, 1975
9. Hardwood forest, North Carolina	6.9	6.2	−0.7	Swank and Douglass, 1977; Johnson and Swank, 1973
10. Oak–pine forest, Long Island	9.7	3.3	−6.4	Woodwell and Whittaker, 1967; Woodwell et al., 1975
11. Beech forest on sandstone, Germany	12.7	12.8	+0.1	Heinrichs and Mayer, 1977
12. Spruce forest on sandstone, Germany	13.5	12.8	−0.7	Heinrichs and Mayer, 1977
13. Northern hardwood forest, New Hampshire	13.9	2.2	−11.7	Likens et al., 1977
14. Tropical rain forest, New Guinea	24.8	0	−24.8	Turvey, 1974
15. Aspen forest, Michigan	19.4 to 38.8	8.3	−11.1 to −30.5	Richardson and Lund, 1975
16. Mixed mesophytic forest, Tennesee	27.4	10.5	−16.9	Shugart et al., 1976
17. Montane tropical rain forest, Puerto Rico	43.1	21.8	−21.3	Jordan et al., 1972
18. Tropical moist forest, Panama	163.2	29.3	−133.9	Golley et al., 1975

26

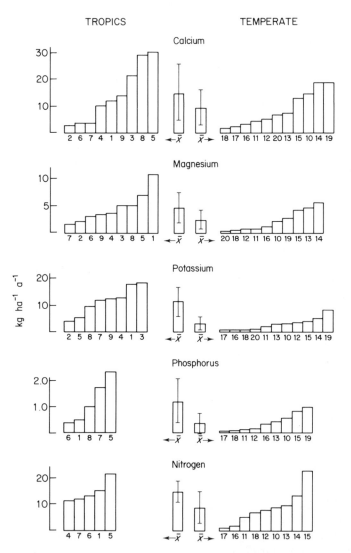

Fig. I.7 Annual amounts of nutrients carried by precipitation into ecosystems in tropical and temperate sites. The number at the bottom of each bar identifies the site. The average annual rate ($\overline{X}$) and standard deviation are given for each nutrient for tropical and temperate regions. The data were obtained from the following areas: (1) Ghana (Nye and Greenland, 1960); (2) N.E. Australia (Brasell and Sinclair, 1983); (3) Puerto Rico (Jordan *et al.*, 1972); (4) Venezuela (Jordan, 1982a, Jordan *et al.*, 1982); (5) Ivory Coast (Bernhard-Reversat, 1975); (6) Amazon, Brazil (Klinge and Fittkau, 1972); (7) Cameroon (Boyer, 1973); (8) Panama (Golley *et al.*, 1975); (9) Malaysia (Kenworthy, 1971); (10) Tennessee (Henderson *et al.*, 1977); (11) North Carolina (Swank and Douglass, 1977); (12) Ontario (Foster and Morrison, 1976); (13) England (Carlisle *et al.*, 1966); (14) Belgium (Duvigneaud and Denaeyer-DeSmet, 1970); (15) Germany (Heinrichs and Mayer, 1977); (16) Oregon (Sollins *et al.*, 1980); (17) Washington (Cole *et al.*, 1967); (18) New Hampshire (Likens *et al.*, 1977); (19) Florida (Ewel *et al.*, 1975); (2) New Mexico (Gosz, 1975).

Aerosol particles originate from a variety of sources: salt from the ocean, dry soil (dust), volcanic ash, organic matter (such as pollen and spores), soot, and industrial haze (Bowen, 1979).

The annual amounts of nutrients carried into 11 temperate and nine tropical ecosystems by wet-fall plus dry-fall are given in Fig. I.7. For all elements, the tropical ecosystems averaged greater yearly input from the atmosphere than temperate ecosystems, but this may be a result of the higher amounts of rain in the tropical sites sampled. Nevertheless, the differences are important because they show that in tropical rain forests, where potential for nutrient loss by leaching is greatest, potential for nutrient restoration through precipitation also may be relatively high.

The difference between atmospheric input of nutrients and runoff losses gives the net change in nutrient stocks during the period of measurement (Table I.2, columns A−R). In ecosystems where total runoff rates were highest (17 and 18), total losses were only partially offset by high rates of atmospheric input, and net loss rates were still relatively high. In contrast, some of the lowland rainforest ecosystems which had the lowest rates of total runoff actually showed a net gain or a very small net loss.

How is it possible that, despite the high potential for nutrient leaching in lowland rain forests on highly weathered soils, actual nutrient loss (as exemplified by calcium) is quite low? The mechanisms for nutrient conservation in these forests are the subject of the next chapter.

E. Chapter summary

This chapter began with a discussion of temperature, rainfall, their seasonality, and how they control nutrient cycling through regulation of such biotic factors as production and decomposition. Because these processes continue year-round in the humid tropics, annual rates of biological activity there are higher than in other regions of the world. High rates of biological activity result in high rates of nutrient uptake by producer and consumer organisms and high rates of release by decomposer organisms. High rates of biological activity also result in high potential for nutrient leaching and weathering of parent material. While the potential for nutrient loss is high in the humid tropics, actual loss from intact, functioning ecosystems often is low and in many cases, nutrient losses apparently do not greatly limit primary productivity. The question of how naturally occurring forests in the humid tropics flourish on nutrient-poor soils and how the forests prevent nutrient loss, despite the high potential for leaching, is the subject of the next chapter.

Chapter II

Nutrient conserving mechanisms

Despite the high potential for nutrient leaching in the humid tropics, there is no evidence that leaching losses from undisturbed forests are any greater, on the average, than losses in temperate regions (Table I.2). How is it possible that these tropical rain forests not only survive, but maintain productivity in spite of the high potential for leaching losses? Many rain forest species have evolved mechanisms that result in very efficient use of the scarce nutrients and that minimize nutrient loss. This chapter describes adaptations which enable native forests to survive in a nutrient-poor environment, and discusses how the adaptations result in nutrient conservation.

Nutrient-poor soils are common in tropical regions. Over half of all tropical soils are highly weathered and leached (Sanchez, 1976). Adaptations to nutrient-poor environments occur in any region where soils are infertile, but because poor soils are common in tropical regions, nutrient conserving mechanisms often are associated with tropical species. Many of the adaptations can be found in nutrient-poor ecosystems throughout the world, and are not confined to the tropics. For example, pine forests growing on impoverished sands in the north temperate zone have adaptations similar to those in some nutrient-poor tropical ecosystems (Jordan and Herrera, 1981).

A. Roots and the below-ground community

Perhaps the most conspicuous adaptations of plants to nutrient-poor soils are the production of a relatively large root biomass (Hermann, 1977) and the concentration of that biomass on or near the soil surface (Jordan and Herrera, 1981). Figure II.1 compares root concentration and distribution in certain nutrient-rich soils (left) and nutrient-poor soils (right) of Europe. The nutrient-poor soils have both more roots, and higher concentrations of roots near the soil surface.

1. Root biomass

The importance of nutrient movement through soil to plant roots in agricultural soils has been discussed by Nye and Tinker (1977). While solute movement

29

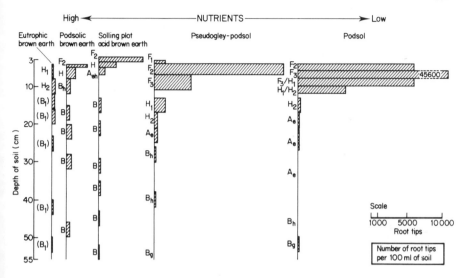

Fig. II.1. Numbers of root tips per 100 cm³ of soil in various European soil types. The soils are arranged in order from the most nutrient rich on the left to the most nutrient poor on the right. F, fermentation layer; H, humus layer; Ae, eluvial horizon; B, illuvial horizon. [Adapted from Meyer and Göttsche (1971) with permission of Springer-Verlag.]

also may play a role in the nutrition of forest species, trees growing on nutrient-poor soils in the humid tropics often have a relatively high root biomass, suggesting that under such conditions a large root system is an important adaptation for acquiring sufficient nutrients.

A high root : shoot ratio may be characteristic of some species genetically adapted to nutrient-poor soils (Chapin, 1980). Other species exhibit a plastic response to nutrient availability. For example, some agricultural species growing on nutrient-poor soil will increase the proportion of photosynthate devoted to root production relative to individuals of the same species on richer soil (Gerloff and Gabelman, 1983).

Forests with high root biomass occupy more fully the volume of soil where nutrients are held after release from decomposition. This means that the average distance nutrients move through the soil to a root is relatively short and the probability of uptake by a root before leaching out of the rooting zone is high. Because of the high leaching potential in the humid tropics, rapid uptake is important for nutrient conservation.

Large root biomass also increases the surface area on which nutrients can be strongly adsorbed. Stark and Jordan (1978) applied radioactive calcium and phosphorus to the top of a root mat covering the floor of an Amazonian rain forest. They found that, while only a small percentage was immediately translocated by the roots, 99.9 percent of the activity applied was retained by the roots through a combination of adsorption and uptake. Only 0.1 percent of the activity leached through the mat of roots.

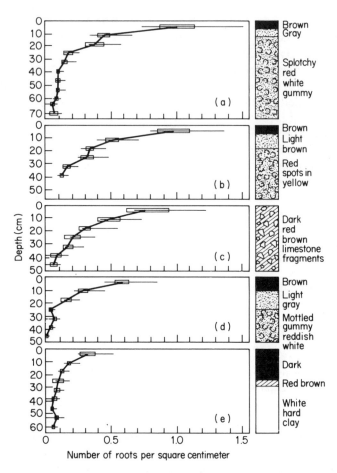

Fig. II.2. Numbers of roots as a function of depth in well-drained soils of various tropical forest ecosystems. Boxes represent one standard deviation. (a) seasonal forest, Barro Colorado Island, Panama; (b) tierra-firme forest, Instituto Agronomia, Belem, Brazil; (c) limestone forest, Cupcake Hills, Puerto Rico; (d) seasonal forest, Fort Clayton, Tropical Test Center, Panama; (e) Cuipo Forest, Bocalara Darien, Panama. [Adapted from Odum (1970a).]

2. Root distribution

Roots in many tropical forests are concentrated near the soil surface (Fig. II.2). High surface concentration may be caused, in some cases, by anaerobic conditions or soil impermeability, but often it appears to be a response to low nutrient availability. For example, in the upper Rio Negro region of the Amazon Basin, where soils are very low in fertility, relatively thick mats of roots occur on top of the soil surface (Stark and Jordan 1978; Jordan, 1985).

When roots are intimately intermixed with litter, decomposers, and other microorganisms which are usually concentrated near the soil surface, the probability of nutrient uptake is increased. It is advantageous to minimize nutrient movement into the mineral soil where nutrients are stored only by exchange on clay surfaces. Nutrients held by exchange are more susceptible to loss than nutrients incorporated in living organisms.

Calculation of the percentage of nutrients entering the root zone that is taken up and recycled by the roots (cycling index; Finn, 1976) showed that the superficial root layer in tropical rain forests can be very effective in recycling nutrients and preventing leaching loss. For nutrient cations in the surface root and humus layer of an Amazonian rain forest (Jordan, 1982b) the ratio is as high as for forests in other regions (between 60 and 80 percent) (Finn, 1978). It is remarkable that the Amazon rain forest, with its high rainfall and acid soil, is able to recycle about the same proportion of nutrients as forests growing in drier regions with less acid soils. The critical factor in the rain forest is the superficial organic matter zone with its high concentration of fine roots and microorganisms.

3. Root–decomposer interactions

The evolutionary significance of concentrating roots near the soil surface is that this can give an individual a competitive advantage for obtaining nutrients released from decomposing litter. A tree that has its roots concentrated near the location of nutrient release, that is, near the decomposers in the soil surface, will have a better opportunity to obtain nutrients than a tree with roots deeper in the soil. Such an advantage is of greater benefit in the most nutrient-poor environments. Equally important, or perhaps more important, is the advantage that an individual tree with roots close to the surface has in competition with the decomposers themselves for the available nutrients.

The interplay between nutrient cycles and populations of trees and decomposers in a nutrient-poor forest could be an example of feedback control of ecosystem function, as follows: In ecosystems on soils low in nutrients, the rate of nutrient release by decomposers may be insufficient to satisfy the nutrient requirements of both the trees and the decomposers themselves. There would thus be competition for nutrients as they become available. The concentration of roots at the site of nutrient release would give trees a competitive advantage, because microbial activity can be inhibited by plant roots (Gosz and Fisher, 1984). However, competitive success by trees would eventually become self-defeating because decomposers require some of the nutrients for their own metabolism. Due to insufficient nutrients, decomposer activity would slow down and undecomposed organic matter would build up. Nutrients immobilized in the undecomposed litter would be unavailable to the trees. This would decrease the growth rate of trees and raise their susceptibility to attack by disease and insects. Eventually, weakened trees would die and fall over, creating a 'gap' in the forest, a phenomenon common in lowland rain forests (Hartshorn, 1978).

Gap formation would be followed by a decrease in living root biomass in the gap area and a decrease in nutrient uptake by roots, with a consequently greater availability of nutrients for decomposers. Gap formation also would increase soil temperature, causing increased decomposer activity and, therefore, an increase in the rate at which nutrients become available.

Gaps in the forest often remain open for a number of years until saplings grow up and fill in the opening (Whitmore, 1978). During the gap filling stage, competition for nutrients in the gap area may be lowered. Nutrients may not severely limit the decomposers. Eventually, however, as the replacement trees get larger, they begin to demand more nutrients. At the same time, their shade reduces the temperature in the decomposition zone on the forest floor. The proportion of available nutrients which the trees appropriate increases and the cycle starts again.

An example of such regulation of nutrient cycling has been described for hemlock forests in Oregon, where slow nitrogen mineralization under closed canopy results in low nitrogen availability for trees (Matson and Waring, 1984). Low nutrient availability results in increased susceptibility to root rot pathogens, which in turn causes death of trees in waves emanating from a central infected tree. The opening of the canopy after die-back results in increased rates of decomposition and nitrogen mineralization, and consequently growth rate of new or remaining trees in the die-back zone is high. The pattern of nitrogen availability is both a consequence and a cause of natural disturbance (Matson and Boone, 1984).

This control of cycling in nutrient-limited forests is an example of cybernetic control of ecosystem function. Engelberg and Boyarsky (1979) have questioned whether ecosystems are cybernetically controlled. They maintained that there is no systematic exchange of information between ecosystem segments and that there are no regulatory or control feedback mechanisms which maintain ecosystems at a fixed set point.

In response, Patten and Odum (1981) argued that a fixed set point maintained by a discrete controlling unit is not necessarily a criterion for the existence of feedback control. They maintained that the interplay of material cycles and energy flows, under informational control, generates self-organizing feedback with no discrete controller required. The switching of nutrient fluxes from trees to decomposers back to trees illustrates how self-organizing feedback might regulate the nutrient cycle without a discrete controller.

4. Aerial roots

The movement of nutrients from decomposing organic matter to roots, without intermediate storage in mineral soil, occurs not only in the root mat on the soil surface, it also occurs within the canopy of the forest itself. Mats of living and dead bryophytes, lichens, club mosses, bromeliads, ferns, orchids, and other epiphytes often occur on the branches and stems of rain forest trees. Adventitous tree roots penetrate the mats. Morphological evidence for the role of such roots in nutrient transfer includes: abundant root hairs, unsuberized root tips, and the presence of endomycorrhizal hyphae (Nadkarni, 1981).

5. Mycorrhizae

It is now well established that the association between plant roots and mycor-rhizal fungi enhances the ability of plants to obtain nutrients under conditions of low nutrient availability (Ruehle and Marx, 1979). Mycorrhizae increase the surface area available for nutrient uptake by roots. In addition, mycorrhizae may also solubilize phosphorus in the soil which otherwise would be unavail-able to plants (Graustein *et al.*, 1977; Sollins *et al.*, 1981). Ectomycorrhizae, so called because they form a sheath of hyphae around rootlets, are very common in the extremely nutrient-poor tropical heath forests (Singer and Silva Araujo, 1979). In other rain forest types, trees often form associations with the vesicular-arbuscular endomycorrhizae which actually penetrate the cortex of feeder roots (St. John, 1980).

Mycorrhizae may provide a 'direct' pathway for nutrients from decomposing litter to roots (Went and Stark, 1968a,b). This pathway bypasses most of the other decomposers on the forest floor such as those illustrated in the detrital food chain in Fig. I.2. The fungal hyphae attach themselves to the decomposing leaves and wood and transfer nutrients to the root (Herrera *et al.*, 1978a). Recent research has shown that mycorrhizal fungi almost completely close tropical rain forest nutrient cycles by their efficient uptake of nutrients from the soil, their advantageous position for scavenging nutrients from dying roots, and their possible interspecific transfer of nutrients (Janos, 1983).

This 'direct cycling' theory has been criticized because it was interpreted by some as suggesting that mycorrhizae were the only, or at least the most important, mechanism of nutrient recycling from litter to roots in tropical rain forests. Clearly, there are many other pathways of nutrient transfer in the forest floor, as indicated in Fig. I.2. The interpretation of 'direct cycling' can be broadened to mean movement of nutrients from decaying organic matter to roots through the manifold pathways depicted in Fig. I.2. This interpretation might be more realistic biologically, and it would be much more meaningful from the view of management and conservation of tropical ecosystems. Con-servation of nutrients is accomplished by their incorporation in living organ-isms, and the objective of management should be to preserve the entire below-ground ecosystem, as much as possible.

The theory of 'direct cycling' also has been criticized because it has been interpreted as indicating that the mycorrhizae actually play a role in the decomposition of litter. In the broadened interpretation of direct cycling, the question would not be important. It is worth noting, however, that there is recent evidence that ectomycorrhizal fungi may in fact have a direct role in litter decomposition (Janos, 1983; St. John and Coleman, 1983).

A mycorrhizal link between litter and root often gives species that have developed this symbiosis a competitive advantage in obtaining nutrients. However, almost all higher plants in all regions of the Earth have evolved some type of mycorrhizal association (Mosse *et al.*, 1981), and there is no evidence that mycorrhizae in the humid tropics are any more effective in recycling nutrients than mycorrhizae of other regions. The lack of relevant evidence is largely due to the difficulty of quantifying the functional role of mycorrhizae.

6. Below-ground community

Nutrient dynamics in the below-ground community are very complex (Witkamp, 1971). During many of the transfers shown in Fig. I.2, a proportion of the nutrients are released in soluble form, and soluble nutrients also may be excreted by many of the organisms. Soluble nutrients may then be adsorbed on the surface of the nearest microorganism, or they may follow the flows indicated in Fig. I.2. They also may be carried downward by percolating water and become adsorbed on clay surfaces in the mineral soil. Nutrients adsorbed on clay surfaces can be taken up by the roots of higher plants, or leached by water percolating through the soil. Percolating water can carry the nutrients down through the soil below the root zone, or laterally to nearby streams.

The below-ground community outlined in Fig. I.2 can be considered a nutrient conserving system. As long as nutrients are bound in the bodies of the various plants and animals, the nutrients are not easily leached or volatilized. Destruction of the below-ground community destroys part of the nutrient-retaining capability of the soil. For example, a comparison of forest management practices to eliminate undesirable seedlings showed that harvesting, removal of slash, and cultivation of the soil resulted in much greater nitrogen volatilization than herbicide treatment only (Vitousek and Matson, 1984). Microbial uptake of nitrogen during the decomposition of residual organic material was the most important process retaining nitrogen in the herbicide treated plots.

The flux of nutrients between soluble forms and forms bound in soil microbes is similar to the spiral of nutrients between soluble and particulate forms in stream ecosystems (Newbold *et al.*, 1982). They proposed a model in which most of the nutrients occurred in the bound form when nutrient limitation in streams is severe, but when nutrient limitation is moderated, nutrients occur principally in the dissolved phase.

B. Leaves

1. Morphology and physiology

Evergreen, scleromorphic leaf types have long been associated with drought-stressed ecosystems such as the Mediterranean vegetation of southern Europe. However, similar leaf types are also associated with nutrient-poor tropical forests (Medina and Klinge, 1983; Medina *et al.*, 1984). Leaves in these forests often are low in nutrient content compared to leaves in more nutrient-rich environments (Peace and Macdonald, 1981).

The young leaves of evergreen tropical species may be susceptible to insect attack, but as they mature, becoming thick and tough, their resistance to herbivory, to fungal infection, and to attack by other predators and parasites increases (Coley, 1982, 1983). Leaves that are long lived, tough, and insect resistant may be advantageous to plants in nutrient-poor areas where leaf replacement is energetically expensive (Chapin, 1980). Leaves are also often protected from herbivory by pubescence, thorns, spines, hairs, and odors (Atsatt and O'Dowd, 1976; Hartshorn, 1978). The leaves of tropical evergreen trees are often more scleromorphic than are those of tropical deciduous trees (Medina and Klinge, 1983) and those of temperate deciduous trees. Werger and Ellenbroek (1978) have shown a decrease in the toughness of leaves along a tropical–temperate gradient. They have suggested that leaves of many tropical species have adopted a 'high-cost, slow profit' strategy to better take advantage of the relatively long growing season. In contrast, short-lived leaves in regions with a short growing season have what they called a 'low-cost, quick profit' strategy. The scleromorphic nature of many rain forest leaves seems to result in lower photosynthetic rates compared to deciduous leaves with higher specific leaf area and higher nitrogen content (Medina and Klinge, 1983).

Another nutrient conserving mechanism characteristic of many species is the retranslocation of nutrients from leaf to twig before leaf abscission (Charley and Richards, 1983). Phosphorus, nitrogen, and potassium are relatively mobile, and are readily retranslocated (Chapin, 1980).

2. Epiphylls

Leaves in the humid tropics are often covered with epiphylls such as mosses, lichens, and algae. Some of the lichens and algae fix nitrogen (Forman, 1975; Whitmore, 1975). Almost all epiphylls appear to be effective in scavenging nutrients from rainwater moving across the leaves. They can be as effective in nutrient removal as a laboratory ion exchange column (Witkamp, 1970; Jordan et al., 1979b). Scavenging also occurs on clean leaves. Fine dust, pollen, and other aerosols that come in contact with leaves can be exchanged or adsorped on leaf surfaces by physical processes (Parker, 1983).

Since nutrients intercepted by leaves eventually reach the forest floor, interception by leaves may not be an important nutrient conserving mechanism for the forest as a whole. However, if at least part of the nutrients intercepted moves into the leaves, the process could give a competitive advantage to the individual trees. The extent to which nutrients intercepted by epiphylls actually move into leaves is not known.

3. Secondary plant compounds

Secondary plant compounds, such as phenolics, may serve as chemical defenses against pathogens and herbivores (Levin, 1976). However, Coley (1983) and Lowman and Box (1983) found that resistance to insect herbivory was better correlated with leaf toughness than with phenol content in the species that they studied.

The concentrations of secondary compounds in plants of nutrient-poor regions has been hypothesized to be high, because of the importance of protection against nutrient loss through herbivory (Janzen, 1974). The hypothesis is based, in part, on the observation that 'blackwater' rivers, colored by these secondary plant compounds, are common in regions of low soil fertility. There is some evidence that in nutrient-poor regions, resistance to herbivory is related to the phenolic content of the vegetation (McKey *et al.*, 1978). However, many tropical species produce secondary compounds regardless of soil fertility. The compounds conspicuously color the stream water only in regions where there is little clay in the subsoil to absorb the compounds (St. John and Anderson, 1982). Absence of clay in the soil may be the factor responsible for both low soil fertility and for the color of the water.

Secondary plant compounds also may inhibit bacterial activity and decomposition, both in living trees and on the forest floor. For example, the leaves of almost all the species examined in an Amazon rain forest near San Carlos de Rio Negro, Venezuela, contained phenols (Sprick, 1979). Denitrifying bacteria could not be detected in soils from that same forest (Gamble *et al.*, 1977).

4. Drip tips

One of the most frequently discussed, yet probably most inconsequential, phenomena in tropical rain forests is the presence of 'drip tips'. Drip tips refer to the acuminate distal ends of leaves of many tropical rain forest species. Dean and Smith (1978) reviewed the argument that it is advantageous to plants to have drip tips if the tips accelerate the drainage of water from the leaf. This might improve plant water relations and also might conserve nutrients by lessening the time that water is in contact with the leaf. Although the experiments of Dean and Smith (1978) showed slower drainage from the leaves of one species from which tips had been excised, G. Parker (personal communication), using another species, found no difference between leaves with excised tips and control leaves. No studies have been done on pine needles, the most hghly developed drip tips.

C. Other mechanisms
1. Bark

The bark of some tree species in nutrient-poor regions is relatively thick. For example, in the upper Rio Negro region of the Amazon Basin, bark constitutes about 10 percent of the trunk weight of the average tree (Jordan and Uhl, 1978). Thick bark may help protect trees from physical injury and subsequent attack by bacteria, fungi, and insects.

2. Soil drainage

The relatively stable structure and large pores of the highly-weathered, iron- and aluminum-rich soils of much of the tropics may be a nutrient conserving mechanism (Nortcliff and Thornes, 1978). The large pores permit rapid drainage and result in diminished opportunity for nutrient exchange between clay surfaces and the drainage water. Consequently, leaching potential is reduced. Although rapid drainage may be a nutrient conserving mechanism, it is a fortuitous one, in contrast to most biological mechanisms which are adaptive.

3. Silicon metabolism

It is well known that many species of tropical trees and grasses contain high levels of silica (Rodin and Bazilevich, 1967; Ludlow, 1969; Koeppen, 1978; Norman, 1979). High silica levels in plants may be related to low phosphate availability which is known to occur in many tropical soils (Sanchez, 1976). One beneficial effect of silicon may be its ability to partially substitute for phosphorus in plants when levels of phosphorus are low (D'hoore and Coulter, 1972).

Another possible reason that silica is accumulated in many tropical species may be the capacity of silicates to replace phosphates bound with iron and aluminum in the soil, thereby releasing the phosphates in soluble form (Taylor, 1961; D'hoore and Coulter, 1972). Since silicates are often rapidly leached in well drained tropical soils (D'hoore and Coulter, 1972), storage of silica in plants could be an important mechanism for ensuring a supply of phosphate. Silicates excreted through roots could solubilize phosphates bound by iron and aluminum in the soil. The benefits of silicate fertilizers in tropical agronomy are well known, especially for tropical grasses such as sugarcane (Fox *et al.*, 1967).

D. Nutrient storage in biomass

It has been suggested that another nutrient conserving mechanism of tropical forests is the storage of nutrients primarily in the biomass where they cannot be leached, in contrast to other regions where nutrients are stored primarily in the soil (Richards, 1952). Cole and Johnson (undated) have pointed out that this is not necessarily true, but strong evidence for or against the hypothesis has never been brought together. The comparisons of nutrient concentrations and total nutrient stocks in Figs II.3 and II.4 are an attempt to answer this question.

The ecosystems for which data were available for comparison of nutrient concentrations were put in four categories. The first is lowland humid forest ecosystems on highly leached soils, where pronounced storage of nutrients in the biomass might be expected. The second group is montane tropical forests, where leaching may be less and the presence of relatively unweathered parent material might be expected to influence storage in the biomass. Temperate forests were placed in two groups, angiosperm forests and gymnosperm

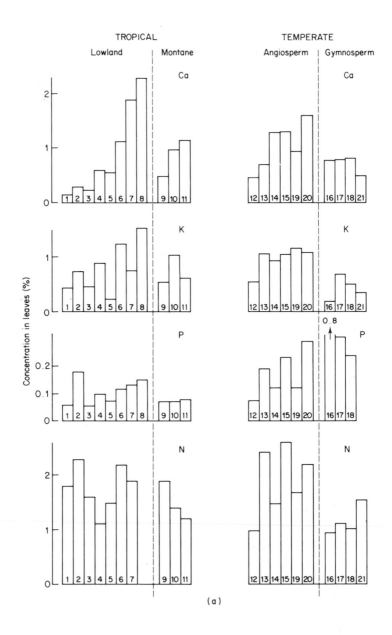

Fig. II.3. Comparisons of concentrations of nutrients in leaves (a) and wood (b) of trees from tropical and temperate forests. The sites and references are: (1) rain forest, Venezuela (Jordan, 1985); (2) rain forest, Manaus, Brazil (Stark, 1971); (3) moist forest, Pará, Brazil (C.E. Russell, 1983); (4) Rain forest, Manaus, Brazil (Herrera *et al.*, 1978b); (5) moist forest, Ivory Coast (Bernhard-Reversat, 1975); (6) 18-year-old-forest, Congo (Bartholomew *et al.*, 1953); (7) moist forest, Ghana (Greenland and Kowal,

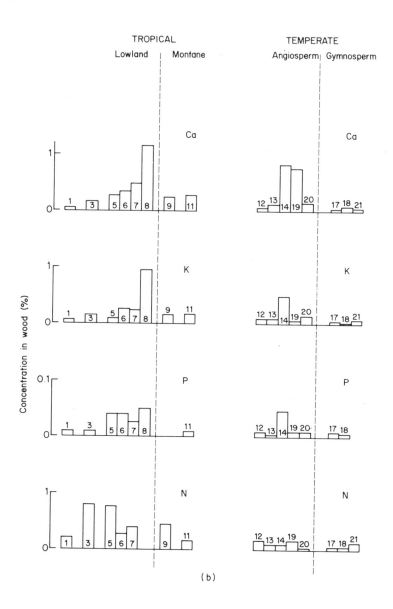

(b)

1960); (8) moist forest, Panama (Golley *et al.*, 1975); (9) slope forest, Colombia (Fölster *et al.*, 1976); (10) montane forest, Jamaica (Tanner, 1977); (11) montane forest, New Guinea (Grubb and Edwards, 1982); (12) oak forest, Long Island, New York, USA (Woodwell *et al.*, 1975); (13) hardwoods, New Hampshire, USA (Whittaker *et al.*, 1979); (14) oak forest, Oklahoma, USA (Johnson and Risser, 1974); (15) deciduous forests, Japan (Tsutsumi, 1971); (16) spruce–fir, British Columbia, Canada (Kimmins and Krumlik, 1976); (17) Douglas fir, Washington, USA (Cole *et al.*, 1967); (18) Douglas fir, Oregon, USA (Sollins *et al.*, 1980); (19) mixed oak, Tennessee, USA (Johnson *et al.*, 1982c); (20) aspen–maple, Wisconsin, USA (Pastor and Bockheim, 1984); (21) balsam fir, New York, USA (Sprugel, 1984).

40

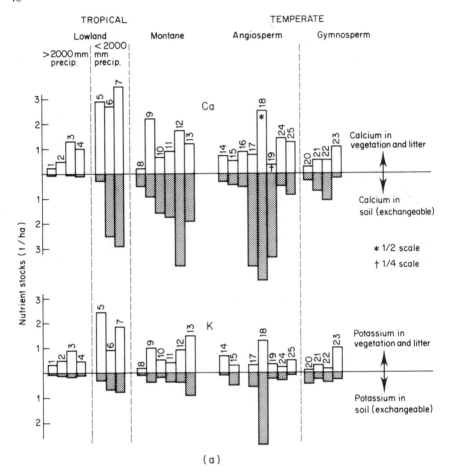

Fig. II.4. Comparisons of stocks of nutrients in biomass (living vegetation plus litter) and soils of tropical and temperate forests. The sites and references are: (1) rain forest, Venezuela (Uhl and Jordan, 1984; Jordan, 1985); (2) rain forest, Manaus, Brazil (Klinge, 1976); (3) moist forest, Pará, Brazil (C.E. Russell, 1983); (4) Banco plateau forest, Ivory Coast (Bernhard-Reversat, 1975); (5) 'old forest', Thailand (Zinke *et al.*, 1978); (6) 40-year-old forest, Ghana (Nye and Greenland, 1960); (7) seasonal forest, Venezuela (Hase and Fölster, 1982); (8) montane forest, Puerto Rico (Jordan *et al.*, 1972); (9) rain forest, Costa Rica (Gessel *et al.*, 1977); (10) montane forest, New Guinea (Manner, 1976); (11) slope forest, Colombia (Fölster *et al.*, 1976); (12) montane forest, New Guinea (Edwards and Grubb, 1982); (13) montane forest, Venezuela (Grimm & Fassbender, 1981); (14) beech forest, Sweden (data of Nihlgard in Cole and Rapp, 1981); (15) beech forest, Germany (data of Ulrich in Cole and Rapp, 1981); (16) hardwoods, New Hampshire, USA (Likens *et al.*, 1977, Bormann *et al.*, 1977); (17) yellow poplar, Tennessee, USA (Henderson *et al.*, 1971; Cole and Rapp, 1981); (18) oak forest, Oklahoma, USA (Johnson and Risser, 1974); (19) oak forest, Virelles, Belgium (Duvigneaud and Denaeyer-DeSmet, 1970); (20) 30-year-old pine forest, Ontario, Canada (Foster and Morrison, 1976); (21) Douglas fir, Washington, USA (Cole *et al.*, 1967); (22) spruce–fir, British Columbia, Canada (Kimmins and Krumlick, 1976); (23) montane fir, Washington, USA (Turner and Singer, 1976); (24) mixed oak,

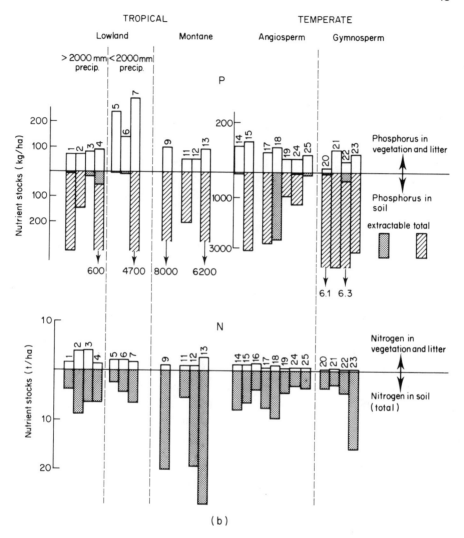

TROPICAL   TEMPERATE

Lowland   Montane   Angiosperm   Gymnosperm

(b)

Tennessee, USA (Johnson *et al.*, 1982c); (25) aspen-maple, Wisconsin, USA (Pastor and Bockheim, 1984).

forests, since this distinction seemed more meaningful than any other division based on the available data.

The concentrations of nutrients in leaf tissue and in wood tissue are given in Fig. II.3(a) and (b). There are no strong and consistent differences in calcium, potassium, and nitrogen concentrations in leaf tissues between the four forest types. For these nutrients in wood tissue, the gymnosperm sites are low. For phosphorus in leaves there are no apparent differences between tropical lowland, tropical montane, and temperate angiosperm systems, but temperate gymnosperms have a higher average concentration. No other strong differences are evident. Thus, it appears that high nutrient concentration in leaf and wood tissue is not a nutrient conserving mechanism characteristic of tropical species.

Another way of examining the question is to compare the nutrient stocks stored in biomass and soil in the tropical and temperate zones (Fig. II.4). The sites were classified in the same four categories as in Fig. II.3, but the lowland tropical category was further divided into sites receiving more than and less than 2000 mm rain per year. In wet lowland tropical sites, there are greater quantities of calcium and potassium stored in the biomass than in the soil. In the drier sites some ecosystems have a larger proportion below ground than do the wet lowland sites. In tropical montane sites, there is a tendency for a greater proportion of the potassium but not calcium to be stored in the biomass. Calcium and potassium in the temperate sites show no strong tendency to be stored either in biomass or in soil.

In comparing the nutrient stocks of the tropical and temperate forests in Fig. II.4, there are more tropical sites with high stocks of calcium and potassium in the biomass than temperate sites. This difference probably is attributable to the disparity in stand age of the tropical and temperate study areas. The temperate sites were generally younger and smaller. For example, site 20 is only 30 years old. For tropical sites with greater than 2000 mm of rain, where nutrient conservation is hypothetically most critical, stocks are not higher than those in temperate forests. Figure II.4(a) provides no evidence that tropical tree species conserve calcium and potassium by putting large stocks in the biomass. Rather, the differences between latitudinal regions is that in the wet lowland tropics these nutrients have been leached from the soil, and therefore, larger *proportions* of the total ecosystem calcium and potassium are stored in the biomass of lowland tropical sites.

There are two other tropical forests for which data on nutrient stocks are available, but which are not included in Fig. II.4 because unusual local circumstances have resulted in extremely high calcium stocks. One is a moist forest in Panama (Golley *et al.*, 1975) located on soils derived from dolomite. The forest had over 4 tons of calcium per hectare in the biomass and over 22 tons per hectare in the soil. The other is a seasonal forest on limestone hills in Costa Rica (Gessel *et al.*, 1977) which had 75 tons of exchangeable calcium and 297 tons of total calcium per hectare.

Figure II.4(b) shows that all natural forest ecosystems in tropical and temperate regions have larger stocks of nitrogen in the soil than in the biomass. None of the forests examined stored and conserved nitrogen primarily in the biomass. The major regional difference illustrated in Fig. II.3 was that nitrogen concentrations were lower in temperate gymnosperm forests. This low concentration is reflected in the low stocks in the above-ground biomass of gymnosperms, shown in Fig. II.4(b).

Phosphorus is the most difficult nutrient to interpret, because some studies reported available soil phosphorus and others total soil phosphorus. Available phosphorus is usually only a small fraction of total phosphorus. This is shown in sites 4, 20,22, and 24 where both were measured. The only striking trend is that the biomass phosphorus stocks of drier tropical forests are relatively high.

The occurrence of high stocks of potassium and phosphorus in the biomass of the drier tropical forests (Fig. II.4(a) and (b)) is not seen in very dry forests, where moisture limits the size of the vegetation. In a dry Puerto Rican forest where annual rainfall is about 900 mm (Lugo *et al.*, 1978), stocks of phosphorus and potassium in the vegetation plus litter were only 5 and 28 kg/ha respectively (Lugo and Murphy, 1984). These values are much lower than those in moister troical forests. The value for nitrogen in the biomass of the dry Puerto Rican forest, 263 kg/ha, also is low compared to the other forests in Fig. II.4(b).

The trend of higher stocks of potassium in the soil of drier forests does appear in the Puerto Rican forest. Total potassium in the soil is 7460 kg/ha, and 25 percent of that is in available form (Lugo and Murphy, 1984). Low leaching is probably partly responsible for the high stocks of potassium in the soil.

In regard to the importance of biomass nutrient storage as a nutrient conserving mechanism in the humid tropics, it appears that for nitrogen and phosphorus it is not important. However, the storage of calcium and potassium in forest biomass could be considered a nutrient conserving mechanism. If the forest is cut, calcium and potassium from decomposing trees is susceptible to rapid leaching, and deforestation could deplete the ecosystem of these nutrients. However, storage of calcium and potassium in the biomass is not an adaptive mechanism in the same sense that a large root biomass is adaptive. Calcium and potassium storage in the biomass of tropical humid forests results simply from the depletion of soil stocks by leaching. It does not result from evolutionary mechanisms that cause nutrients to be hoarded in the biomass.

From the perspective of conservation and management, the most important point about the nutrient conserving mechanisms discussed in this chapter is that they are integral parts of undisturbed, native forests. When such forests are cleared and the sites used for agriculture, forestry, pastures, or other reasons, the mechanisms which conserve nutrients are destroyed. The extent to which these mechanisms are affected, and the ability of the disturbed ecosystem to recover, depend on the nature of the disturbance. Chapter V discusses the changes in nutrient cycles when tropical forests are disturbed.

E. Chapter summary

This chapter addressed the question: Why, despite the high potential for nutrient leaching in the humid tropics, are actual nutrient losses from undisturbed forests apparently low? The answer was, that a variety of nutrient conserving mechanisms reduce nutrient loss. Some of the mechanisms are: large root biomass, concentration of roots near or on the soil surface, mycorrhizae, maintenance of a complex below-ground community, long-lived and resistant leaves, thick bark, rapid soil drainage, and possibly silicon metabolism. Because nutrient storage in biomass has been proposed as a nutrient conserving mechanism, this possibility was also examined. However, the high proportion of total ecosystem calcium and potassium in the biomass of rain forest trees results not from unusually high storage by the trees, but rather from the low nutrient content of the soils.

Perhaps the most significant aspect of all the nutrient conserving mechanisms is that they are all part of the undisturbed, living native forest. When the forest is subjected to natural or human disturbance, some, or all, of the mechanisms may be destroyed.

Chapter III

Differences in ecosystem characteristics along environmental gradients

Not all the adaptations, which enable the forests of the humid tropics to survive and function despite the high potential for nutrient loss, are present in all tropical forests. For example, there are tropical forests in which there is no above-ground root mat, nor even a conspicuous concentration of roots in the upper soil horizon. The adaptations which exist differ both quantitatively and qualitatively in different regions and can be different in stands of trees only a short distance apart.

Adaptations differ along gradients of fertility, temperature, and moisture. Often, many nutrient conserving mechanisms are present and conspicuous in ecosystems on nutrient-poor soil. In ecosystems on nutrient-rich soils, some of the adaptations may be absent or less conspicuous. Adaptations also differ along altitudinal gradients. Here, the changes may reflect changes in temperature. Along a wet to dry moisture gradient, leaching and soil weathering become less severe, and vegetation structure and function are increasingly adapted to shortages of water rather than nutrients.

This chapter examines how the structure and function of ecosystems change along gradients of fertility, altitude, and moisture. Before beginning, however, it is appropriate to present a brief overview of the climate, land forms, vegetation, and soils of the tropics as a background against which the gradients can be discussed.

A. World patterns of land form, climate, vegetation, and soil

Technically, the tropics comprise that area of the world between the Tropic of Cancer (latitude $23\frac{1}{2}°$ N) and the Tropic of Capricorn (latitude $23\frac{1}{2}°$ S). Sometimes it is more practical to use temperature as a basis for the delineation of the tropical zone. For example, the mean annual 20°C isotherm has been used to delimit the tropics (Fig. III.1). Other definitions are: areas with

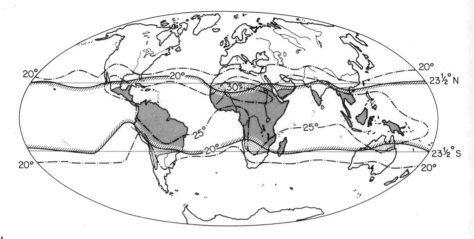

Fig. III.1. Thermal delimitation of the tropics. The mean annual isotherms of 20 °C, 25 °C, and 30 °C are shown by lines of alternating dashes and dots. The 20 °C January (northern hemisphere) and July (southern hemisphere) isotherms, corresponding to the coldest months, are shown by solid lines with hachures. Areas where the mean daily temperature range exceeds the mean annual temperature range are indicated by the dark shading inside the 20° isotherm and dots outside of the 20° isotherm. [Adapted from Tricart (1972) with permission of Longman.]

temperatures which exceed 20 °C during the coldest months of the year and areas that have a mean daily temperature range greater than the mean annual range (Fig. III.1).

In the tropics, as in other regions, climate and geology interact with living organisms to form the natural patterns of vegetation and soils. Holdridge (1967) and Walter (1971) have presented detailed descriptions of the interrelationships between climate and vegetation, and the text accompanying the world soils map of the Food and Agricultural Organization (FAO, 1974) includes detailed descriptions of geomorphology, vegetation, climate, and soils. These maps and descriptions are useful in understanding patterns at a local or regional level, but they are too complex for use on a global scale. To illustrate more clearly the relationships of land form, climatic region, and vegetation on a global scale, greatly simplified maps (Fig III.2 (a)–(c)) have been prepared from the FAO (1974) maps.

The uppermost map in each figure shows the land form, which is a way of greatly simplifying the geology and geomorphology of a region. Interior lowlands are an important feature in the Amazon region of South America and in the central and west-central regions of Africa. These lowlands are surrounded, in part, by higher plateaus. Plateaus also are dominant features in much of South-east Asia and most of Australia. Mountains dominate the landscape on many of the islands of South-East Asia and the Caribbean, along the coasts of India, the eastern coast of Australia, and the highlands of East Africa. The Andes Mountains, running along the western coast of South America, are the largest range of mountains in the tropical zone. The Himalayan Mountains, to the north of India, are outside the tropical region. Deserts are important in northern Africa, the western fringe of South America, and in northwestern India. The term desert is sometimes applied to dry plateaus, as in interior Australia and northern Mexico.

Climate is shown in the middle map of each figure. Interior regions of South America and Africa, and some coastal lowlands often have moist, continually humid climates. These climates are characterized by more than 100 mm of rain each month. In the plateau regions, climate tends to be seasonal or monsoonal. In these areas, up to half or more of the year can lack significant rain. Seasonal climates often grade into semi-arid and dry climates across large plateaus, as in eastern Brazil, Australia, and northern Africa. Semi-arid climates are characterized by low humidity and only scattered rainfall.

The bottom map of each figure represents vegetation. The most highly developed tropical rain forests occur in the center of the tropical lowlands, where the climate is humid year-round. Towards the edge of the lowland areas, adjacent to the plateaus, rain forest grades into moist forest. Moist forests receive less annual rainfall than do rain forests (Holdridge, 1967). In moist forests, some of the trees may shed all of their leaves for a few weeks or months each year, but the forest is never totally leafless. In the plateau regions, where dry seasons are longer, moist forest gives way to seasonal or monsoonal forest which can lose most of its leaves for a portion of each year. In the drier savannas and deserts, plant species may be deciduous or may possess drought-resistant scleromorphic leaves. Savanna grasses have stems and leaves that die during the dry season, but the rhizomes remain alive and sprout again at the start of the rainy season. Montane vegetation is complex and depends on the temperature and moisture regime at each elevation. Montane forests resemble lowland rain forests where rainfall is heavy, but at drier sites can resemble savanna or desert ecosystems.

Since soils are formed by the interaction of climate, geology, and vegetation, soil types are highly correlated with these factors (Table III.1). A discussion of soil types often is confusing because there are several soil classification systems in use. For example, the highly weathered soils occurring under tropical rain and moist forests and capable of forming laterite are classified as 'latosols', in some older systems of classification. In the new FAO (Food and Agriculture Organization) system they are called 'ferralsols', and in the US Department of Agriculture classification they are 'oxisols'. Furthermore, many of the soil classification categories in one system do not have precise equivalents in other systems. The problem is especially difficult in the US classification system in which soil chemical and physical properties rather than soil genesis govern the system. To prevent misrepresentation, in this book, whenever a particular study is discussed, the author's original classification is given.

Table III.1

Relationships between tropical regions, naturally occurring vegetation, and soils (three systems of classification): taxonomic equivalents adapted from Aubert and Tavernier (1972), Brady (1974), Jackson (1964), Sanchez (1976), and FAO–UNESCO (1974)

Region	Vegetation	Soil characterization	Great soil group	FAO classification	US Department of Agriculture classification
Lowland wet tropics	Tropical rain forest, high species diversity	Intensely weathered, silica leached, iron and aluminum oxides' dominant	Latosol	Ferralsol	Oxisol
Lowland moist tropics and subtropics	Rain forest and seasonal forest	Highly weathered, but clay minerals common	Red-yellow podzolic	Acrisol	Ultisol
Areas where rock high in bases is present	Depends on rainfall regime	High in bases	Terra Roxa	Luvisol	Alfisol
Lowland wet tropics	'Tropical heath forest'; 'kerangas'; 'Amazon caatinga'	Intensely weathered, iron and aluminum leached, silica dominant	Podzol	Podzol	Spodosol
Moist and wet mountain regions, volcanically derived	Montane forest, high species diversity	Derived from volcanic materials	Andosol	Andosol	Andept subgroup
Mountain valley	Forest or grassland	Rich in organic matter	Chestnut	Kastanozems	Mollisols
Montane region, non-volcanic	Variable, depending on temperature and moisture	Bedrock shallow to deep, depending on climate and erosion	Lithosol	Lithosol	Lithic subgroup
River banks with annual flooding, i.e. Amazon 'várzea'	Flooding interval determines vegetation	Alluvial deposit on river banks	Alluvial	Fluvisol	Fluvent subgroup
Tropical desert and savanna	Desert and savanna	Little chemical weathering, high in bases	Desert, sierozem	Yermosols	Aridisol
Monsoonal	Monsoon adapted	Swelling-type clays	Grumusol	Vertisol	Vertisol

B. Gradients within regions

Although the broad vegetation classifications in Fig. III.2 imply homogeneity of ecosystems throughout a region, there is actually a high diversity of ecosystems within each region. For example, Fig. III.3 illustrates the diversity of ecosystems which occur at San Carlos de Río Negro, in the Amazon Territory of Venezuela, along a transect of a few hundred meters. The relative proportions of clay and sand in the soil, the soil depth, and the depth of water table appear to determine the type of vegetation. Where the mineral soil above the underlying clay is pure sand, vegetation is small, as in the various caatinga forests. Their low stature could be a result of the low nutrient holding capacity of the sand, occasionally droughty or generally high water table conditions, or a combination of both factors (Klinge and Herrera, 1978; Sobrado and Medina, 1980; Anderson, 1981). The height and biomass are greatest where the sand and clay are mixed and form a deep soil horizon. Forests in this zone are usually dominated by one or a few tree species, often legumes (Buschbacher, 1984). Where the clay forms a distinct and resistant barrier to root penetration at a shallow depth, forest size is intermediate, as in the mixed species (no strong dominants) type of tierra firme forest. In the river-bank forests (igapó), seasonal flooding influences vegetation structure (Kuo Keel and Prance, 1979).

C. Nutrient gradient

Comparison of tropical forest ecosystems within the humid lowland tropical region (Fig. III.2) has been carried out (Jordan, 1985) using the centered and standardized principal components analysis techniques of ordination (Gauch, 1977; Gauch et al., 1977). Because relatively few tropical forest ecosystems have been studied it was necessary to include sites that are perhaps only marginally within the lowland tropical rain forest biome. However, all the sites in the ordination are lowland or lower montane, late successional or primary, continually wet or evergreen seasonal tropical forests (Tables III.2, III.3, III.4)). The following were specifically excluded: forests at high altitude, swamps, mangroves, deciduous forests, early secondary forests, savannas, grasslands, deserts, conifer, and plantation forests.

Two sites are located close to San Carlos de Río Negro, in the Amazon Territory of Venezuela. One site is a heath forest, or Amazon 'caatinga', growing on a coarse sandy spodosol. It is a forest type distinct from the caatinga of the Brazilian cerrado (Anderson, 1981). Other types of heath forest in the Amazon region are the 'campina', which is intermediate in structure, and the 'bana' which is a very reduced forest (Klinge and Medina, 1979). In South-East Asia, heath-type forests are known as 'kerangas' (Brunig, 1974). The second site is a mixed upland forest on an oxisol. The third site is a lower montane forest at El Verde Puerto Rico at an elevation of 500 m. The 'Banco' forest, in the Ivory Coast, is an evergreen forest with two dry seasons per year. The

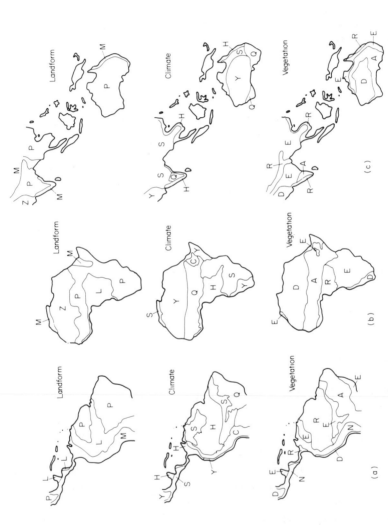

Fig. III.2. Landforms, climates, and vegetation types in tropical America (a), Africa (b), and South-east Asia and Australia (c). Relationships on islands are on a scale too small to indicate. Map key: *Land forms:* L, lowland; P, plateau; Z, desert; M, mountain. *Climate:* H, humid; S, seasonal or monsoonal; Q, semi-arid; Y, dry; C, cool. *Vegetation:* R, rain forest; E, seasonal or monsoonal forest; A, savanna; D, desert; N, montane.

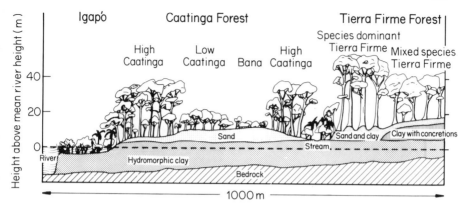

Fig. III.3. A schematic representation of the relation between hydrology, vegetation, and soils in the upper Rio Negro region of the Amazon Basin. Areas close to rivers and streams contain hydromorphic gley soils and support igapó (seasonally flooded) forest. Moving away from rivers the vegetation changes to either caatinga or tierra firme forest. Where surface soils are sandy, igapó grades into caatinga. Caatinga forest areas do not experience annual flooding and soils are classified as spodosols. 'High caatinga' occurs where the sand is fine grained and the water table remains close to (within 100 cm of) the soil surface throughout the year; 'low caatinga' and 'bana' occur on somewhat more-elevated, better-drained sites containing coarse sands and which experience alternating flood and drought (Klinge and Herrera, 1978). Where the clay content of the surface soil is higher and oxic concretions are present in the lower soil horizons, tierra firme forest occurs, either as a legume-dominated forest on ultisols or mixed-dominant stands on oxisols.

'Pasoh' forest in Malaysia is characterized by very tall trees of the family Dipterocarpaceae. The forest at La Selva, Costa Rica is on deep soils of alluvial and volcanic origin. The Panamanian forest is on soils derived from rock high in calcium.

Table II.5 shows the ranking of the ecosystems, with a relative score from 100 to 0. The ranking was not correlated with precipitation, elevation, average temperature, nor actual evapotranspiration at the sites (Table III.2). It also was not correlated with exchangeable calcium, total calcium, soil pH, nor total phosphorus in the soil (Table III.6). However, for the five sites for which total soil nitrogen is available (Table III.6), there is a relationship between score and total soil nitrogen. Assuming an exponential relationship between the score (Table III.5) and the soil nitrogen, the coefficient of determination (r^2) is 0.94. This suggests that soil nitrogen may be the significant environmental factor represented by the first axis. However, an examination of all the available ecosystem parameters along the gradient (Table III.3) gives further insights about the ranking.

Root biomass (parameter 1, Table III.3) decreases regularly in the ranking from Caatinga to Panama forest, except for an inversion of the San Carlos oxisol and El Verde montane sites. Relatively high root biomass can be taken

Table III.2

Locations of the ecosystems used in ordination: sources of data are listed in Table 3.4 unless otherwise indicated

	Latitude and longitude	Location	Precipitation (mm/year)	Approximate meters above sea leval	Average temperature (°C)	Actual evapo-transpiration (mm/year)
San Carlos, Venezuela*	01°56'N 67°03'W	Near the common border of Venezuela, Colombia and Brazil	3565	100	26	1778
El Verde, Puerto Rico	18°19'N 65°45'W	Luquillo mountains, eastern Puerto Rico	2920	500	23	1752
Banco Forest, Ivory Coast	5°N 4°W	Near Abidjan	2095	100	26	1530†
Pasoh, Malaysia	2°58'N 102°17'E	140 km southeast of Kuala Lumpur	2054	100	26	1515†
La Selva, Costa Rica	10°N 84°W	Foot of central mountains, northeast Costa Rica	4300	40	24‡	2153‡
Darien Province, Panama	8°38'N 78°08'W	Sante Fe	2000	<250	25	1442†

* Two ecosystem sites were near San Carlos.
† From Meentemeyer (undated).
‡ From Luvall (1984).

Ecosystem characteristics in tropical moist and rain forests: numbered references and footnotes are given in Table 3.4

Parameter	Amazon caatinga San Carlos, Venezuela	Oxisol forest San Carlos, Venezuela	Lower montane rain forest El Verde, Puerto Rico	Evergreen forest Banco, Ivory Coast	Dipterocarp forest Pasoh, Malaysia	Lowland rain forest La Selva, Costa Rica	Moist forest, Panama
1. Root biomass (t/ha)	132[1]	56[2]	72.3[7]	49[17]	20.5[19]	14.4[14]	11.2[23]
2. Aboveground biomass (t/ha)	268[1]	264[2]	228[8]	513[17]	475[21]	382[15]	326[23,25]
3. Root : shoot ratio	0.49	0.21	0.32	0.10	0.04	0.04	0.03
4. Root distribution, percentage in superficial root mat (%)	26[1]	20[2]	~0[7]	~0	~0[19]	~0[14]	—
5. Specific leaf area (cm²/g)	47[3]	65[4]	61[7]	—	88[21]	139[26]	131–187[23,25]
6. Leaf area index	5.1[1]	6.4[4]	6.6[9]	—	7.3[21]	—	10.6–22.4[23,25]
7. Predicted leaf biomass (t/ha)	10.8	9.8	10.8	—	8.3	—	10.4
8. Leaf litter production (t ha^{-1} a^{-1})	4.95[2]	5.87[2]	5.47[8]	8.19[17]	6.30[22]	7.83[16]	11.3[23]
9. Predicted turnover time of leaves (years) (row 7 ÷ row 8)	2.2	1.7	2.0	—	1.3	—	0.9
10. Aboveground wood productivity (t ha^{-1} a^{-1})	3.93[2]	4.93[2]	4.86[8]	4.0[17]	6.4[21]	—	—
11. Leaf decomposition, k	0.76[2]	0.52[2]	2.74[10]	3.3[17]	3.3[22]	3.47[16]	3.2[23]
12. Biomass : phosphorus ratio, leaf litter fall	2631[1]	7237[10]	5000[6]	1365[18]	3282[20]	2024[26]	1319[23]
13. Biomass : nitrogen ratio, leaf litter fall	135[1]	95[2]	—	64[18]	82[20]	52[27]	—

Table III.4
References and footnotes for Tables III.2, III.3 and III.6

1. Herrera, 1979	15. Werner, 1982
2. Jordan, 1985	16. Gessel *et al.*, 1977
3. Medina *et al.*, 1978	17. Lemee, undated
4. Jordan and Uhl, 1978	18. Bernhard-Reversat, 1975
5. Jordan and Herrera, 1981	19. Yoda, 1978
6. Luse, 1970	20. Lim, 1978
7. Odum, 1970b	21. Kato *et al.*, 1978
8. Jordan, 1971	22. Ogawa, 1978
9. Jordan, 1969	23. Golley *et al.*, 1975
10. Wiegert, 1970	24. Average of seven plots
11. Edmisten, 1970	25. Average of two sites
12. Jordan *et al.*, 1972	26. Luvall and Parker, unpublished data
13. Bourgeois *et al.*, 1972	27. Cole and Johnson, undated
14. Raich, 1980a	

Table III.5
First axis ranking of ecosystems in Table III.3 according to centered and standardized principle components analysis: the first axis accounted for 63.2 percent of the variance among the ecosystems; subsequent axes accounted for 17 percent or less, and probably were ecologically meaningless (Jordan, 1985)

Ecosystem	Score
San Carlos, spodosol	100
San Carlos, oxisol	71.5
El Verde	45.8
Banco	15.7
Pasoh	13.4
La Selva	3.2
Panama	0

as an indicator of nutrient-poor sites when flooding or subsoil hardpan are not present (Chapin, 1980). Above-ground biomass (parameter 2) does not show a strong trend along the gradient, but the root:shoot ratio (parameter 3) parallels the root biomass gradient. The presence of roots in a mat above the soil surface (parameter 4) occurred only in the San Carlos ecosystems, although individual roots often occur above the mineral soil in other sites. Roots close to the surface may be a selective advantage in obtaining nutrients from decomposing litter in areas where soil nutrient availability is low.

Specific leaf area (parameter 5), the ratio of the leaf area to weight, follows the same trend from Caatinga to Panama, with only San Carlos oxisol and El Verde montane inverted. Leaves that have a low area:weight ratio can be indicative of stressful environments (Grubb, 1977). Thick leaves are often called scleromorphic or sclerophytic. Along a fertility gradient in the Brazilian cerrado region, scleromorphic features of the vegetation were positively correlated with low levels of extractable phosphate in the soil (Goodland and

Table III.6

Nutrient characteristics of the soils at the ecosystem comparison sites: numbered references are given in Table III.4

Parameter	Amazon caatinga San Carlos, Venezuela	Oxisol forest San Carlos, Venezuela	Lower montane rain forest El Verde, Puerto Rico	Evergreen forest Banco, Ivory Coast	Dipterocarp forest Pasoh, Malaysia	Lowland rain forest La Selva, Costa Rica	Moist forest, Panama
Exchangeable calcium in A_1 soil horizon (mEq/100 g)	0.57[1]	0.03[2]	1–5[5]	0.09[18]	0.13[19,24]	1.28[13]	36[23]
Total calcium in soil (kg/ha)	195[2]	7[2]	176[12]	—	115[19,24]	6530[16]	22 166[23]
Soil pH	4.0[1]	3.9[2]	4.3–4.8[11]	4.1[18]	4.3–4.8[23]	4.0[13]	4.7–5.9
Total nitrogen in soil (kg/ha)	785[1]	1697[2]	—	6500[18]	6752[19,24]	20 000[16]	—
Total phosphorus in soil	36[1]	243[2]	—	600[18]	44[19,24]	7000[16]	22[23]

Pollard, 1973). In Jamaica (Loveless, 1961, 1962) and in Australia (Beadle, 1962, 1966), low soil phosphate also was strongly and positively correlated with scleromorphic features of the vegetation.

Leaf area index (parameter 6) increases regularly from Caatinga to Panama. If specific leaf area and leaf area index are considered simultaneously, another interesting trend emerges. The weight per unit area of leaves (one over the specific leaf area) times the leaf area index times an areal correction factor predicts the total leaf biomass of each site. When this is done (parameter 7), no trend is apparent, and the sites do not seem to differ from one another in quantity of leaves. The uniformity of predicted leaf biomass is striking. The reason may be that where leaf area index is low, the leaves are relatively thick (low specific leaf area), while where leaf area index is high, the leaves are thin.

Leaf litter production generally increases from San Carlos to Panama, but the increase is not regular (parameter 8). However, when leaf biomass (parameter 7) is divided by leaf production (parameter 8) to obtain the turnover time of leaves, the ratio along the gradient decreases almost regularly (parameter 9). If long-lived leaves are an indication of stress, as indicated by Chapin (1980), the ordination suggests that the ecosystem ranking may be caused by stress.

Because data are lacking, there is no clear trend in wood production rates (parameter 10). The decay constant for leaves (parameter 11) is lowest in the two San Carlos forests, indicating that decomposition there is slowest. The slow decay is probably attributable to the same factors that cause the toughness and longevity of the leaves. The constant k increases regularly in the ordination, except for a slightly lower value at the Panama site.

Parameter 12 shows the biomass:phosphorus ratio in freshly fallen litter. Biomass:nutrient ratios (or their inverse, nutrient concentration) measured in freshly fallen litter may be a convenient parameter for indicating nutrient use efficiency by the vegetation at a particular site (Vitousek, 1982). Comparisons of nutrient concentrations and nutrient use efficiencies can suggest which element is limiting (Leech and Kim, 1981; Vitousek, 1984). High nutrient use efficiency (low concentration) of a particular nutrient suggests that that nutrient is relatively scarce at a site. Excepting the Caatinga and Banco sites, biomass:phosphorus ratios decrease regularly along the gradient. This suggests that low phosphorus availability may be a stress in some of the tropical forest ecosystems, and may play a role in establishing the gradient shown by ordination. Vitousek (1982, 1984), on the basis of nutrient concentrations in freshly fallen litter from diverse worldwide forests, suggested that nitrogen was more often limiting in temperate ecosystems, and that phosphorus was more often limiting in tropical ecosystems.

The Caatinga site has a high biomass:nitrogen ratio (parameter 13), suggesting that, in contrast to the other sites, nitrogen is a critical nutrient there. Another indication that nitrgoen is limiting in the Caatinga site is the frequent occurrence there of carnivorous plants (e.g. *Drosera* spp.; Klinge and Medina, 1979). It is commonly suppose that plant carnivory is an adaptive response to low levels of soil nitrogen (Heslop-Harrison, 1978), although aquisition of other nutrients also may be important (Lüttge, 1983).

The comparisons of the parameters suggest that no single parameter alone is responsible for the environmental gradient represented by the first axis of ordination. Phosphorus may be one of the important parameters. Despite the lack of correlation of scores with total soil phosphorus, the strong gradient of foliar phosphorus concentration, excepting the Caatinga site, suggests that 'available', rather than total soil phosphorus, may be critical. 'Available' phosphorus is difficult to quantify. In the laboratory, 'availability' is determined by the solubility of phosphorus in an extracting solution. Under field conditions, plants may differ widely in their ability to take up phosphorus from the same soil. These differences can be related to root uptake kinetics, type of mycorrhizae, and other factors (Chapin, 1980).

The high correlation of score with total soil nitrogen suggested that the latter may have been an important factor represented by the first axis of ordination. Total nitrogren is linked to plant growth through the rate at which the stocks of nitrogen in the soil become mineralized (Swift et $al.$, 1979). Therefore, the litter decay coefficient, k, may be a better indicator of nitrogen importance than total stocks of soil nitrogen. Leaf decomposition constants (Table III.3, parameter 11) suggest that nitrogen may be most critical in San Carlos sites.

Soil water conditions may also distinguish the Caatinga site from the other sites. The deep coarse sands of the Caatinga do not hold moisture effectively in the root zone. Water stress can occur within a few days after a rainstorm, and adaptation to both low nutrient levels and drought may cause the sclerophylly (Sobrado and Medina, 1980).

The soil types at the seven sites implicate soil nutrient status as an important factor in the ecosystem rankings. The Caatinga forest is on a spodosol, an extremely nutrient-deficient soil type (Sanchez 1981). The other San Carlos site, as well as the Banco site (Lemee, undated) and most of the Pasoh site (Yoda, 1978), are on oxisols or ultisols. These soils also are usually acid and deficient in plant nutrients (Sanchez et $al.$, 1982). In contrast, the soil at La Selva is relatively high in most nutrients due to its volcanic origin (Bourgeois et $al.$, 1972), and at the Panama site the soil is relatively rich because of underlying dolomite (Golley et $al.$, 1975). Only the El Verde site, situated on volcanic bedrock (Little, 1970) seems out of place along a gradient based on soil type. Perhaps the higher elevation of the El Verde site is important in its relative ranking, or perhaps the great age of the volcanic soils (50 million years; Little, 1970) is implicated.

It would be convenient to have simple, descriptive terms for the aggregate characteristics of the contrasting types of ecosystems. Whitmore (1975) has used the term 'oligotrophic' to characterize the heath-like forests of South-East Asia, which are structurally similar to the Caatinga forest, and which also occur on podzolized sands. The terms 'oligotrophic' and 'eutrophic' have limnological connotations. Nevertheless, rather than coin new terms, it might be appropriate to refer to ecosystems at the Caatinga end of the gradient as oligotrohpic, and to those at the Panama–La Selva end of the gradient as eutrophic.

1. The Sarawak nutrient gradient study

The ordination (Table III.5) suggested that a gradient of soil fertility was important in the ranking of the seven rain forest ecosystems. It is unlikely that climatic differences strongly influenced this ranking. However, to evaluate the influence of soil fertility on the structure and function of ecosystems, it is preferable to compare ecosystems having the same climate. The influence of climate was probably negligible in a series of studies comparing four tropical forest ecosystems on a single island. The forests, on four different soil types, and all located within a 10 km radius, were studied in Gunung Mulu National Park, Sarawak (Proctor *et al.*, 1983a,b; Anderson *et al.*, 1983). The study sites were: heath forest on podzol soil, Dipterocarp forest on red-yellow podzolic soil, forest on river-deposited alluvium, and forest on limestone derived soil.

The heath forest is comparable to the Amazon Caatinga on spodosol (Table III.3). The dipterocarp forest is comparable to the Pasoh forest on oxisol. The forest on alluvium probably is most similar to the La Selva site of those forests in Table III.3. The limestone forest is comparable to the Panama forest.

Using the rankings of the analagous forests (in Table III.3), the Gunung Mulu forests were aligned from oligotrophic to eutrophic in Table III.7. As does the Caatinga (in Table III.3), the heath forest in Gunung Mulu has the highest biomass:nitrogen ratio (row 1); that is, it has the lowest nitrogen concentration in the leaf litter. This suggests that in each listing, the heath forests are most critically limited by nitrogen. Although total soil nitrogen at the heath forest site (row 3) is relatively high, the low litter decay factors (rows 9 and 10) suggest that nitrogen is mineralized slowly, and this could cause low foliar nitrogen concentration.

Phosphorus in the leaf litter at Gunung Mulu (row 2) follows the same pattern as at the sites in Table III.3. After the heath site, where nitrogen appears to be more critical, biomass:phosphorus ratios are highest in the forest on oxisol, lower in the alluvial forest, and lowest in the limestone forest. Since phosphorus availability depends on available calcium and soil pH (Brady, 1974), the increasing calcium (row 5) and increasing pH (row 6) along the oligotrophic–eutrophic gradient suggest that available soil phosphorus should similarly increase from the dipterocarp forest to the limestone forest in Table III.7. If it does, it could explain the higher phosphorus concentration in the litter of the eutrophic sites. Total soil phosphorus (row 4) is unrelated to the gradient, but total phosphorus has no consistent relationship to available phosphorus.

Above-ground biomass in the Sarawak study (Table III.7, row 7) has little relationship to the oligotrophic–eutrophic ranking but leaf litter fall (row 8) is higher in the eutrophic sites. Both these parameters follow the same trends as in the oligotrophic–eutrophic ranking in Table III.3.

Table III.7

Characteristics of rain forests along nutrient gradient in Sarawak (from Proctor et al., 1983a,b, Anderson et al., 1983)

| | Oligotrophic— — — — — — — — — — — — — — — — — —Eutrophic | | | |
	Heath forest	Dipterocarp forest	Alluvial forest	Limestone forest
1. Biomass : N in leaf litter	176	105	111	85
2. Biomass : P in leaf litter	7042	9524	3745	2660
3. Total N in soil (t/ha)	7800	6000	7800	5000
4. Total P in soil (t/ha)	190	360	420	120
5. Exchangeable Ca in soil (t/ha)	62	4.6	1600	2400
6. Soil pH	3.6	4.1	4.4	6.1
7. Above-ground biomass (t/ha dry weight)	470	650	250	380
8. Leaf litter fall (t/ha^{-1} a^{-1})	5.6	5.4	6.6	7.3
9. Decay constant, k, for leaves	1.4	1.7	1.7	1.7
10. Decay constant, k, for total small litter*	1.3	1.3	1.7	1.5

* Leaves, twigs, flowers, fruit.

60

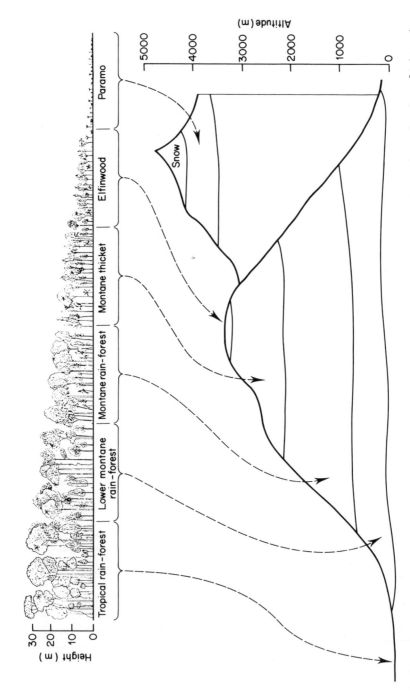

Fig. III.4 Profile diagram of an ecocline along an elevational gradient on tropical mountains in continental regions. [Adapted from Whittaker (1975) with permission of Macmillan Publishing Co.]

The decay constants (k) for leaf litter and total small litter (rows 9 and 10) generally follow the oligotrophic–eutrophic gradient, with higher decay rates and constants characterizing the more eutrophic ecosystems. The trends are not as strong as might be expected, and the authors attribute this to factors such as moisture differences between sites (Anderson et al., 1983).

Although the absolute values of some ecosystem parameters at the Gunung Mulu sites were different than those at comparable sites in Table III.3, most of th parameters followed the same relative trends along the oligotrophic–eutrophic gradient. The most notable exception is total stocks of soil nitrogen. Total soil nitrogen at the Sarawak heath site is as high as that at other Sarawak sites, and much higher than the Amazon Caatinga site. This suggests that the nitrogen mineralization rate rather than the total stock of nitrogen, is more critical.

The similarity between the trends in the Sarawak study, where climatic variation was known to be small, and the forests listed in Table III.5 suggests that climatic differences probably are relatively unimportant in the ranking of the forests. A consideration of all the variables suggests that the rate at which soil nutrients become available probably is the most important factor.

D. Altitudinal gradient

The structure and function of tropical forest ecosystems changes along an oligotrophic–eutrophic nutrient gradient. The structure and function of ecosystems also change markedly along an altitudinal transect from lowlands to mountain tops.

One of the first scientists to write about the differences in ecosystems along an altitudinal gradient in the tropics was Alexander von Humboldt, following his explorations of the Ecuadorian Andes in 1802. The following is an excerpt from an account of these explorations (E. L. Jordan, 1981):

> At first, Humboldt's mountaineering excursions – he scaled in successive trips of two to three weeks the volcanos Pichincha, Cotopaxi, Antisana, and Illinica – were largely geographical and geological outings. But then he discovered something the jungle wilderness of the Orinoco had not disclosed. He saw the Andes as a vertical exhibit of the earth's climates, plants, and animals, from hot tropical lowlands to the eternal snow and ice of the towering summits, which at that time were considered the highest roof of the world. The slopes disclosed a distinct number of strata, each life zone consisting of a definite grouping of soils, land forms, rainfall, temperature, plant species, and animal species. These environments – today we would call them ecosystems – had not assembled themselves accidentally or arbitrarily but had grown as communities with specific characteristics determined by interaction. All component forces continuously acted on each other and created a balanced unit.

Figure III.4 illustrates the change in structure of vegetation with altitude on a tropical wet mountain. Descriptions of the forests at different elevations have been given by Beard (1944) and Brown et al. (1983) among others. The lowland rain forest at the foot of the mountain contains trees of all heights and diameters, including some 35 m or more in height and a meter or more in diameter. Canopy leaves, although thicker than the leaves of most temperate forests, are still relatively thin, and although sometimes covered with epiphylls, the coverage usually is not heavy. Except in lowland oligotrophic forests, there usually is not a large accumulation of litter and humus.

As one climbs the mountain to the lower montane forest, there are fewer really big trees, and the height of the canopy declines to 20–30 m. The species composition changes: sometimes gradually, sometimes abruptly. Canopy leaves are a little thicker and more heavily covered with epiphylls. In surface depressions, there may be pockets of accumulated undecomposed organic matter.

Further up the mountain, there is another transition, sometimes fairly sharp, to the upper montane rain forest. Here the canopy is lower still, sometimes less than 20 m, and almost all the leaves are laden with epiphylls. There is a noticeable accumulation of organic matter on the soil surface.

The montane forest may grade directly into an upper montane forest, variously called sub-alpine rain forest, cloud forest, dwarf forest, or elfinwood forest. Here the stature of the forest is very reduced. Sometimes the canopy is only a few meters high and the biggest trees are often less than 15 cm in diameter, but the trees are densely crowded so that walking through the forest is difficult. The canopy is heavily covered with thick layers of epiphylls. Mosses, lichens, and algae also cover exposed rocks. The soil surface is covered with a thick layer of undecomposed organic matter often heavily penetrated by roots. The forest is frequently covered by clouds which reduce transpiration, and the condensation on leaves increases humidity and soil moisture (Weaver et al., 1973). Wind often is another stress on high elevation plant communities.

Sometimes, along the transition between montane rain forest and elfin forest, there occur thickets of one or a few plant species apparently adapted to unique conditions. For example, in the transition between upper montane rain forest and cloud forest in the Luquillo mountains of Puerto Rico, slopes are often dominated by a single species of mountain palm in colonies called 'palm brakes' (Beard, 1949).

Where tropical mountains are sufficiently high that frost occurs, there may be a distinctive vegetation called páramo (Walter, 1971). Páramos resemble deserts in structure, even though the environment may be wet Many species are sclerophyllous, and the vegetation is sparse and open. On the highest peaks, páramo grades into the region of continual snow.

On many tropical islands and peninsulas, where the mountains rise abruptly from sea level, the lowest elevations may not be covered with tropical rain forest, but instead with seasonal and dry-evergreen forests (Fig. III.5). At mid-elevation there is a transition to montane rain forests, and above that the vegetation communities are similar to those in Fig. III.4.

Why does the altitudinal gradient in vegetation exist? The gradient is produced by several interacting factors. Warm, moisture-laden air moves

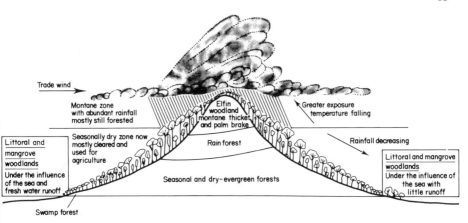

Fig. III.5. Profile diagram of an ecocline along an elevational gradient on mountains on small tropical islands. [Adapted from Beard (1949).]

across the surface of the ocean in tropical latitudes (Fig. III.5). When the air encounters a mountain, it rises and, as a consequence, cools. Cooling causes an increase in humidity and, at higher altitudes, the moisture condenses and forms clouds (Rumney, 1968). Precipitation increases with altitude and causes the transition from dry forest to montane rain forest on islands (Fig. III.5). At the higher altitudes, the continual cloud cover and high humidity slows down evapotranspiration and can result in waterlogged soils (Grubb, 1977).

The presence of a large land mass influences the altitude at which cloud forests occur in the tropics. On small tropical islands, cloud forests often occur at less than 2000 m, while on large land masses, which exert a greater heating and drying effect on the incoming oceanic air, cloud forests occur above 3500 m. The compression of life zones on small land masses, as compared to continents, is called the 'Massenerhebung effect' (Grubb, 1977).

1. Effect of gradient on ecosystem function

A series of studies in montane forests in Jamaica and New Guinea (Table III.8) suggests that there is a gradient of nutrient availability along altitudinal gradients on tropical mountains. A comparison of these montane forests with the forests along the oligotrophic–eutrophic gradient in Table III.3 shows how stress acts similarly along both altitudinal and nutrient gradients. The above-ground biomass in the New Guinea forest (Table III.8, row 2) is similar to that of the lowland rain forests in Table III.3 (row 2), while the Jamaican forest biomass is lower, and comparable only to the lower montane rain forest of Puerto Rico. The root:shoot biomass ratios (Table III.8, row 4) show that the New Guinea forest is comparable to forests intermediate along the oligotrophic–eutrophic gradient (Table III.3, row 4), while the Jamaican forest is comparable to the oligotrophic sites. Leaf litter fall (Table III.8, row 5)

Table III.8

Ecosystem characteristics of montane rain forest on a large land mass (New Guinea: Edwards, 1977, 1982; Edwards and Grubb, 1977, 1982; Grubb and Edwards, 1982) and of rain forest on a small land mass (Jamaica: Tanner, 1977, 1980a,b)

	New Guinea	Jamaica (Mor Ridge)*
1. Altitude (m)	2500	1600
2. Above-ground biomass (t/ha)	310	223
3. Below-ground biomass (t/ha)	40	54
4. Root shoot	0.13	0.24
5. Leaf litter fall (t/ha^{-1} a^{-1})	6.4	4.9
6. Height of canopy (m)	33–37	5–7
7. Biomass : N (leaf litter)	72	—
8. Biomass : P (leaf litter)	1333	—
9. Biomass : N (mature leaves)	76	95
10. Biomass : P (mature leaves)	1162	2000
11. N in soil (kg/ha)	54 000	9000
12. Specific leaf area (cm^2/g)	58	58
13. Leaf area index	4.9–6.0	4.1

* 'Mor' is a term for a type of forest humas layer of inincorporated organic material, usually matted or compacted, and distinct from the mineral soil (Brady, 1974).

follows the same trend, the New Guinea forest being comparable to the more eutrophic sites and the Jamaican forest to the oligotrophic sites. The canopy height of the two montane forests (Table III.8, row 6) differ greatly, the Jamaican site being almost a dwarf forest.

The biomass:nutrient ratios of leaf litter are not available for the Jamaican site (Table III.8, rows 7 and 8). For the purpose of comparison with these values in Table III.3, biomass:nutrient ratios in mature leaves (Table III.8, rows 9 and 10) will be used. In the New Guinea forest, the difference between ratios in leaves and litter are not large.

The biomass:nitrogen ratio of the New Guinea forest (Table III.8, row 9) is intermediate along the fertility gradient (Table III.3, row 13). The biomass:nitrogen ratio in the Jamaican forest leaves is exceeded only by that of the Caatinga forest, even though soil nitrogen at the montane sites (Table III.8, row 11) is higher than that at most of the lowland forests (Table III.6). Because foliar nitrogen concentrations can be low in montane tropical forests, in spite of high soil nitrogen concentrations, organic matter decomposition and nitrogen mineralization may be important rate-limiting steps in the functioning of the high altitude tropical forests.

The biomass:phosphorus ratio in the leaves of the Jamaican and New Guinea forests (Table III.8, row 10) are at the low end of the range of litter values in Table III.3, row 12. This indicates that phosphorus availability is not as critical as nitrogen availability at these montane sites.

The specific leaf area of both montane forests (Table III.8, row 12) was 58 cm^2/g, a value lower than those of all the rain forests along the oligotrophic–eutrophic gradient in Table III.3, row 5, except the Caatinga forest. The leaves of the montane forests are, thus, relatively sclerophyllous. Grubb (1977) has called the thick leaves of cloud forests 'pachyphylls' to distinguish them from the 'sclerophylls' of Mediterranean type vegetation.

Leaf area index values also suggest similarities between the montane forests and the most oligotrophic forests. Leaf area index in the Jamaican Mor Ridge (Table III.8, row 13) is lower than any of the lowland rain forests (Table III.3, row 6), and the index of the New Guinea forest also is relatively low.

Because of the 'Massenerhebung effect', the gradient of vegetation characteristic on mountains is not strictly an effect of elevation, but rather of climate which is affected by both elevation and size of land mass. Thus more extreme effects are found at the Jamaican site than at the New Guinea site, even though the latter is at a higher elevation. Despite difficulties of comparison due to this effect, it seems that the changes in environment along a mountain slope have an effect on nutrient availability and vegetation structure and function similar to that along the gradient of increasingly weathered soils. Both the most oligotrophic lowland forest (Caatinga) and the most extreme montane forest (Jamaica, Mor Ridge) are similar in many respects. Both are characterized by low biomass, low height, low productivity, low leaf area index, slow decomposition, low nitrogen concentration in biomass and litter, high surface litter accumulations, accumulation of roots on the soil surface, high root:shoot ratio, and low specific leaf area. A major difference between the lowland heath forests and the cloud forests is the much heavier epiphyll cover in the cloud forests, probably resulting from the higher average humidity at the higher elevations.

Studies of soil organic carbon and nitrogen along altidudinal gradients in Colombia (Alexander and Pichott, 1979) and in Thailand (Yoda and Kira, 1969) also suggest that decreasing rates of decomposition, as a function of altitude, may cause nitrogen deficiency in high montane forests. High amounts of soil carbon, and a high carbon:nitrogen ratio suggest low rates of decomposition (Swift *et al.*, 1979). In the Colombian study, soil carbon increased from 6.9 kg/m^2 at 600 m altitude to 33.8 kg/m^2 at 3700 m, and carbon:nitrogen ratios increased from less than 10 to 15 or more. In Thailand, total organic carbon increased from about 5 kg/m^2 at sea level to nearly 15 kg/m^2 at higher altitude, while the carbon:nitrogen ratio also increased from approximately 10 to 15.

Low decomposition at high elevation largely results from the lower temperatures and lower solar radiation caused by heavy cloud cover (Grubb, 1977). Soil pH also decreased with elevation at the Colombian and Thailand sites, but low pH could be either an effect or a cause of slow decomposition.

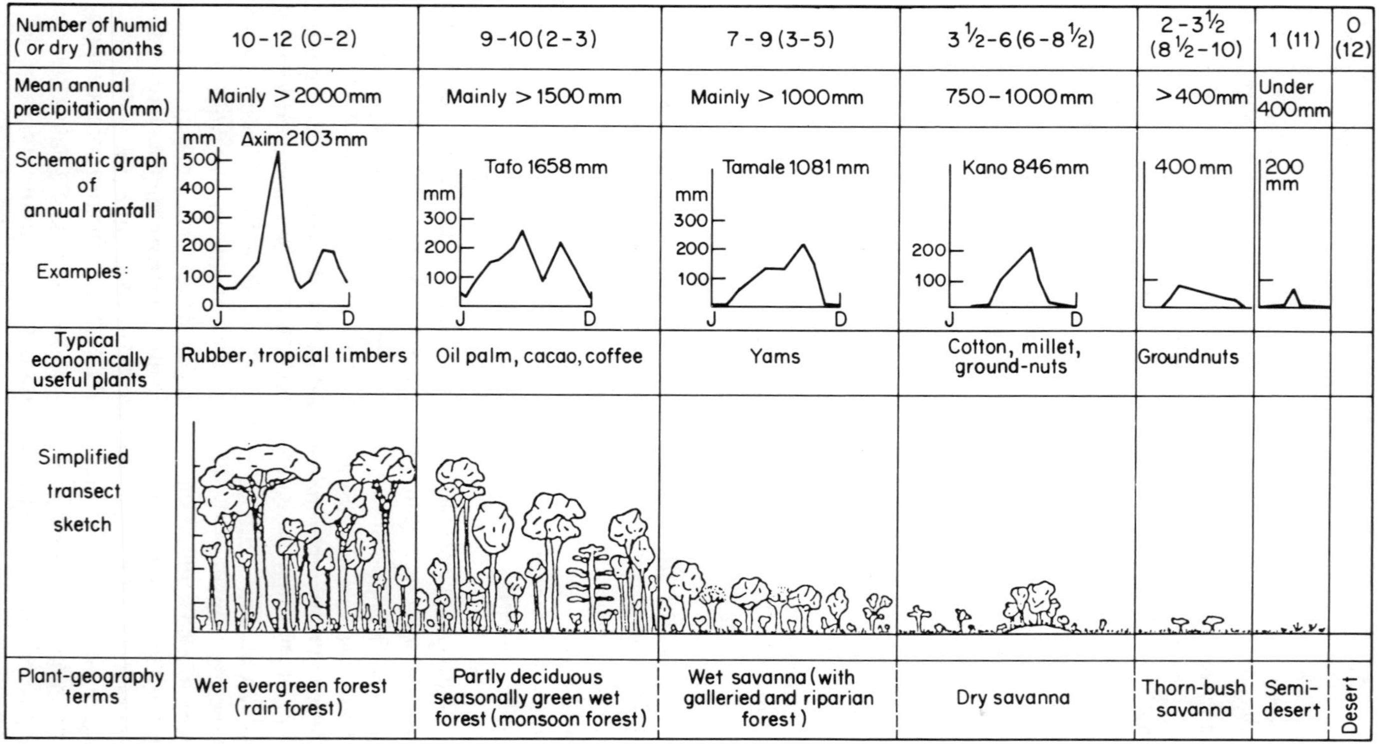

66

Number of humid (or dry) months	10–12 (0–2)	9–10 (2–3)	7–9 (3–5)	3½–6 (6–8½)	2–3½ (8½–10)	1 (11)	0 (12)
Mean annual precipitation (mm)	Mainly > 2000 mm	Mainly > 1500 mm	Mainly > 1000 mm	750–1000 mm	> 400 mm	Under 400 mm	
Schematic graph of annual rainfall	Axim 2103 mm	Tafo 1658 mm	Tamale 1081 mm	Kano 846 mm	400 mm	200 mm	
Typical economically useful plants	Rubber, tropical timbers	Oil palm, cacao, coffee	Yams	Cotton, millet, ground-nuts	Groundnuts		
Simplified transect sketch							
Plant-geography terms	Wet evergreen forest (rain forest)	Partly deciduous seasonally green wet forest (monsoon forest)	Wet savanna (with galleried and riparian forest)	Dry savanna	Thorn-bush savanna	Semi-desert	Desert

Fig. III.6 Schematic summary of tropical vegetation formations along a moisture gradient and examples of major crops that are common in each region. [Adapted from Manshard (1974), with permission of Bibliographisches Institut AG Mannheim.]

2. Mid-elevation tropical valleys

In some mid-elevation tropical valleys, temperatures are intermediate between those of the lowlands and the upper montane forests. Often, this intermediate temperature is combined with a seasonal rainfall pattern. As a result, the leaching potential is lower than that of lowland wet forests, but decomposition rates are higher than those of upper montane forests. In addition, alluvium from higher mountain slopes often improves the fertility of the valley soils.

Such valleys sometimes are intensively cultivated because they provide favorable conditions for plant growth. In Central and South America, these valleys were the favored locations for Spanish settlement. Many of these colonial cities were, and still are, important agricultural centers. However, the high crop productivity achieved in these valleys is atypical of the tropics in general, and results obtained from agricultural experiment stations in such locations are not often applicable to other areas.

E. Moisture gradient

The structure and function of tropical forest ecosystems change in a predictable manner along nutrient and altitudinal gradients. Changes in ecosystem structure and function also occur along a moisture gradient from rain forest through savanna to desert (Fig. III.6). For example, biomass, height, and leaf area index are low, and leaf sclerophylly is more developed in dry Mediterranean-type ecosystems (Margaris, 1981) and tropical savannas (Sarmiento and Monasterio, 1983), compared to rain forests. In desert regions, the development of scleromorphic forms of vegetation is extreme (Solbrig and Orians, 1977). Vegetation is sparse, and leaves are thick, with small cells, thick walls, and thick cuticles.

The most obvious environmental variable along a regional moisture gradient is annual precipitation, but precipitation alone is not the best indicator of water stress, since the seasonal distribution of rainfall greatly influences vegetation development. A better index of moisture stress is the ratio of potential evapotranspiration to precipitation (Holdridge, 1967). Potential evapotranspiration is the rate at which evaporation could take place from an open water surface if such a surface were present. Thus, where precipitation is evenly spread throughout the year, potential evapotranspiration can be lower than at sites with highly seasonal rainfall, because there are more rainy days during the year. Rainy days have higher humidity than dry days and, consequently, lower potential evapotranspiration.

Along a gradient from humid to dry, potential evapotranspiration increases and precipitation decreases. Thus, the ratio between the two increases. Figure III.7 shows above-ground net primary productivity as a function of the ratio at 10 North American sites where data were collected as part of the US International Biological Program. The relationship between moisture stress and primary productivity is linear, on a log–log scale, along the gradient through

68

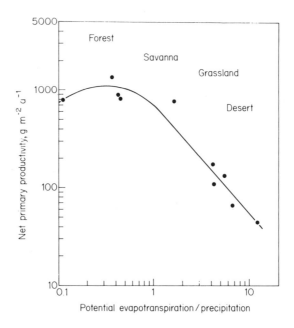

Fig. III.7. Above-ground net primary productivity as a function of the ratio between potential evapotranspiration and precipitation for 10 sites in the United States. Points are calculated from the data tabulated by Webb *et al.* (1983). Temperate data are used here because comparable tropical data are not easily available.

desert, grassland, and savanna ecosystems (right to left in Fig. III.7). However, between dry forest and wet forest, factors other than moisture become limiting and productivity declines as the ratio of evapotranspiration to precipitation approaches 0.1.

Along the gradient from humid to dry ecosystems, the potential for leaching and weathering decreases, partly because less water percolates through the soil. In ecosystems where potential evapotranspiration exceeds precipitation, the long-term net direction of water movement in the soil is upward. Another reason for decreased leaching and chemical weathering in drier areas is that net primary productivity is reduced. As explained in Chapter I, when productivity is low, decomposition and soil respiration are also low and the production of carbonic acid is decreased.

Lower leaching results in larger soil stocks of calcium and potassium in dry, as compared to wet, tropical sites (Fig. II.4(a)). Larger soil stocks of these cations probably are responsible for the larger stocks of alkaline cations found in the biomass of forests on the drier sites.

Alkaline cations also influence phosphorus availability in the soil through the regulation of soil pH (Brady, 1974). The nearly neutral soil pH in some of the drier forests may account for the higher stocks of phosphorus in their biomass (Fig. II.4(b)). However, the absence of leaching and chemical weathering in dry regions does not necessarily mean that nutrients are always abundant. Other factors, such as geological history and the nature of the parent rock, are also important. For example, in dry areas of Australia, phosphorus, nitrogen, and some of the trace elements are deficient (Williams and Colwell, 1977).

1. Saline and alkaline soils

Overabundance of nutrient elements can sometimes be a problem in arid areas. For example, when a water table is relatively close to the soil surface, the predominantly upward movement of the soil solution carries dissolved elements towards the surface (Peck, 1977). Due to evaporation from the soil surface, salt accumulation can occur in the upper soil layers. The more mobile ions, such as sodium, potassium, calcium, magnesium, chloride, sulfate, carbonate, and bicarbonate, are those most frequent in the zone of salt accumulation (Williams and Colwell, 1977).

Cutting of forests in dry regions can cause increased salinity at the soil surface. Decreased transpiration results in more water percolating into the soil. The water table rises and with it the zone of salt accumulation (Peck, 1977). The problem is likely to occur more often in valleys, following the removal of forests on the slopes.

Even in arid and semi-arid areas where there is no groundwater table nor irrigation, accumulations of calcium and magnesium carbonate (caliche) and calcium sulfate (gypsum) can occur in the upper soil horizons. There is often a significant eolian deposition of calcium on the soil (Gile *et al.*, 1966). Calcium accumulations in the soil also may occur as a result of carbonate eluviation from surface horizons in arid zone soils (Arkley, 1963). In arid regions where the soil does not become saturated, and deep drainage does not occur, carbonates are precipitated in lower soil horizons as soil water evaporates or is transpired (Jenny, 1980).

F. Pine forests in the tropics

In the humid temperate zone, forests dominated by *Pinus* spp. often occur on nutrient-deficient soils, such as highly leached, coarse podsols (Forman, 1979). Pines may be successful in such habitats because their sclerophyllous needles act to conserve nutrients, being long-lived and resistant to leaching, insect attack, and decay (Monk, 1966). Undecomposed organic matter often accumulates on the soil surface. This organic matter, sometimes called 'duff' or 'mor' is often heavily penetrated by roots. The roots of pines usually have ectomycorrhizal associations which greatly increase their efficiency of nutrient recycling. Temperate coniferous forests can be very conservative of nutrients (Stone and Kszystyniak, 1977). The drainage water from the sandy soils is clear but dark in color, and such streams are often called 'blackwater' streams (Jordan and Herrera, 1981).

Tropical analogs of the temperate pine forests on podsols are the Amazonian caatingas and the kerangas of South-East Asia. Like the temperate pine forests, these forests have sclerophyllous leaves, a thick mat of roots and partially decomposed humus on the soil surface, ecotomycorrhizal associations, and dark but clear drainage water (Jordan and Herrera, 1981).

Forests dominated by the pine also occur in various regions of the tropics. However, tropical pine-dominated ecosystems appear to be functionally analagous to the coniferous forests of the Northwestern United States on relatively fertile soils (Sollins et al., 1980) rather than to the temperate coniferous forests on nutrient-poor spodosols (McFee and Stone, 1965). In the Pacific Northwest region, domination by conifers may be related to seasonality (Waring and Franklin, 1979). When temperatures are optimum for growth during the summer months, moisture is very low. Precipitation occurs mostly during the winter when temperatures are sub-optimal for growth. Because most conifers keep their leaves year-round, they are better able than deciduous species to take advantage of precipitation when it occurs.

Although the evergreen trees of the Pacific Northwest appear to have overcome, at least to some extent, the environmental limitations posed by dry summers and wet winters, decomposers apparently have been less successful. These environmental limitations may cause the relatively low decomposition rates found in these forests (Olson, 1963). Because of the low annual decomposition rates, nitrogen mineralization is slow, little nitrogen is available for uptake by roots, and the nitrogen content of living trees is low, despite average levels of total soil nitrogen soil (Fig. II.4(b)). Nitrogen, in a form available for uptake by roots, appears to be the factor limiting growth (Sollins et al., 1980).

There apparently have been no studies on the factors influencing the occurrance of pines in tropical regions. However, naturally occurring pine forests in Mexico, Guatemala, Honduras, and South Vietnam occur where a hot and dry season alternates with a cooler and moister season suggesting that as in the Pacific Northwest, nitrogen mineralization is a critical factor in determining the type of forest.

Pines are capable of growing in seasonally dry tropical savannas where available nitrogen is scarce if mycorrhizae are present, and plantations of pines in lowland tropical areas are becoming increasingly common (Lamb, 1968). However, the pines there do not seem to readily reproduce by seed and, when replanted after a first rotation harvest, frequent weeding is required to prevent competition from native broad-leaved species (personal observations, Jarí plantation, Pará, Brazil, and plantations of the Corporación Venezolana de Guayana near the Orinoco Delta, Venezuela).

G. Generalized stress gradient

Many of the characteristics typical of the stressed environments discussed in this chapter were similar despite the nature of the stress. Ecosystems in both nutrient- and moisture-stressed environments exhibited relatively low stature, low biomass, and low net primary productivity. Analyses of ecosystems subjected to other types of stress have revealed similar responses. For example, Woodwell (1967, 1970) has analyzed ecosystem response to radiation stress and compared it to the effects of stress by pollution, salt spray, exposure on mountains, and water availability. In general, the most highly stressed ecosystems showed smaller biomass, lower species diversity, lower net primary productivity, and a lower photosynthesis:respiration ratio.

It appears that an important mechanism causing the generalized stress response is increased respiration relative to photosynthesis. For example, higher respiration rates may occur in polluted ecosystems due to the need for detoxification. High root respiration in nutrient-poor terrestrial ecosystems may reflect the relatively greater root production necessary to acquire essential nutrients.

Increased respiration is a probable explanation for the absence of large trees in stressed ecosystems. Large trees have a large proportion of non-photosynthetic biomass supported by a relatively small mass of leaves. For this reason shrubs are generally more stress tolerant than trees, grasses more tolerant than shrubs, and herbs more tolerant than grasses. Under conditions of very high stress, the only plants that generally survive are those which consist almost entirely of photosynthetic cells, such as algae.

The changes in ecosystem characteristics along a gradient of increasing stress are the reverse of some of the changes observed during the process of succession. During the transition from early to late successional stages, biomass, diversity, and size of organisms increase (Odum, 1969). Succession appears to occur after a relaxation of stressful conditions, and represents a reversion of ecosystem structure to pre-disturbance conditions.

H. Chapter summary

In this chapter the characteristics of naturally occurring ecosystems along three major environmental gradients were examined. The first was a gradient correlated with the state of weathering of the soil in evergreen forested ecosystems. The ecosystems on the most highly weathered soils (spodosols, oxisols) displayed the highest development of nutrient conserving mechanisms. Because nutrients were strongly implicated, it was proposed that the forests characteristic of the two ends of the gradient be termed oligotrophic and eutrophic.

The second was an elevational gradient. The changes in ecosystem characteristics along an altitudinal transect paralleled those along the eutrophic–oligotrophic gradient. Evidence suggested that the rate of nutrient mineralization from decomposing organic matter caused high elevation ecosystems to resemble those suffering nutrient deficiency. Exceptions to the elevational gradient occur in locations where seasonality or topography reduce precipitation and humidity for part, or all of the year.

The third was a moisture gradient. Some of the characteristics of moisture-stressed ecosystems were similar to those of nutrient-stressed ecosystems. A review of ecosystem response to stress in general suggested that ecosystems respond in similar ways, regardless of the nature of stress.

Chapter IV

Characterization of nutrient cycles

Evidence showing that nutrients can be critical in tropical rain forest environments, and that native species have adapted in various ways to nutrient shortage, has not been extensively reviewed, and a major objective of this book is to develop that evidence. In contrast, comparisons between the cycles of individual elements and the processes which govern those cycles have been extensively reviewed. Despite the many recent publications on nutrient cycling processes, it is worthwhile to briefly review this literature, as a basis on which to build the subsequent chapters of this book. These chapters will deal with the effects of perturbation on nutrient cycles in tropical ecosystems.

The number of chemical elements found to be essential for life keeps increasing as experimental methods improve (Frieden, 1972; Miller and Neathery, 1977). However, the cycles of all nutrient elements can be classified into one of two different types (Deevey, 1970). In the first, the nutrient element occurs in a volatile form during part of the cycle. Nitrogen and sulfur are two nutrient elements which follow this pattern. Carbon, hydrogen, and oxygen, which are the major building blocks of living protoplasm, also occur in volatile form during part of their cycles. Although these latter three elements are constituents of organic matter, they are not generally dealt with in discussions of nutrient cycling and soil fertility, but rather in the context of energy and water flux. The other type of nutrient cycle is the non-volatile cycle, characteristic of all other nutrient elements.

A. Non-volatile cycle

Although the rates of flow of non-volatile nutrient elements through the ecosystem differ, as do their stocks in given ecosystem compartments, the pathways and storage compartments of all non-volatile nutrients are similar (Fig. IV.1).

Nutrients can enter an ecosystem through the weathering of minerals in the subsoil. Where parent rock is relatively shallow, as in mountainous areas, mineral weathering may be an important source of some nutrients. In other

73

74

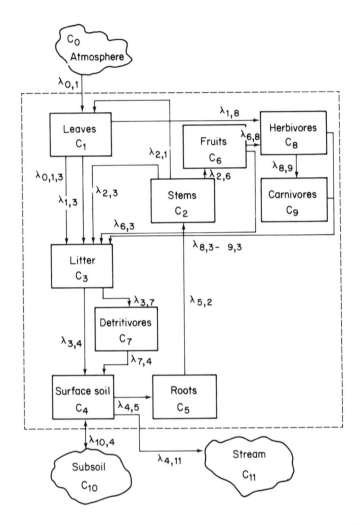

Fig. IV.1. A diagrammatic model of a tropical forest ecosystem. The diagram depicts nine organic compartments within the ecosystem and illustrates the flow of non-volatile elements through the system. Transfers between compartments are identified by λ, with subscripts indicating the compartment of origin and the compartment of destination. Forest compartments are represented by rectangles, compartments external to the forest system are represented by amorphous figures. [Adapted from Golley *et al.* (1975) with permission of the University of Georgia Press.]

areas, such as the central Amazon Basin where there has been no geological uplift for millions of years, minerals in the subsoil may be completely weathered and depleted of nutrients (Kronberg *et al.*, 1979) and the underlying parent rock may be deeply buried beneath an impermeable layer of saprolite or highly weathered clay (Jenny, 1980). In such areas, the underlying rock does not constitute an important source of nutrients (Jordan, 1982a).

Nutrients also enter ecosystems through atmospheric deposition of aerosols and fine particulates such as dust and pollen (Manokaran, 1980). The aerosols can settle out due to the force of gravity, a process called sedimentation, or they can be deposited on leaves by wind, a process called impaction. Both sedimentation and impaction are classified as 'dry fall' or 'dry deposition' (Swank, 1984). If the aerosols are carried into the ecosystem dissolved or suspended in rainwater, the process is called precipitation scavenging.

In temperate ecosystems, some of the nutrients entering with rainfall are washed through the canopy and into the soil where they are available for uptake by roots. In tropical humid ecosystems, where older leaves are often heavily covered with epiphylls, some of the nutrients in the rainfall may be scavenged by the epiphylls as the rain passes over the leaves (Witkamp, 1970).

Precipitation which passes through the canopy is called 'throughfall', and precipitation which flows down the trunks of trees is called 'stemflow'. Usually, the concentration of dissolved nutrients in throughfall is higher than that in precipitation, due to the leaching of nutrients out of leaves by the rain (Parker, 1983). However, in a nutrient-poor rain forest with a high load of epiphylls, concentrations of some nutrients were lower in the throughfall than in the precipitation (Jordan et al., 1979b).

When dissolved nutrient ions in throughfall and stemflow enter the soil they may replace ions already held by the charged surfaces of clay or humus particles (colloids) through a process called exchange. Clay and humus colloids have negative surface charges which hold positively charged ions (cations) such as calcium, potassium, hydrogen, and aluminum (Brady, 1974). Although one species of cation may replace another already held by a clay particle, there is no single universal order of replacement (Jackson, 1964). However, there are two important generalizations with regard to cation replacement. One is that monovalent cations, such as potassium and sodium, are not as strongly held as divalent cations, such as calcium, and are more easily replaced by hydrogen ions (Wiklander, 1964). The second generalization is that replacement depends, in part, on the relative activity of the cations in the soil solution and those adsorped on the soil particles. Thus, when large amounts of calcium (e.g. in ash from forest burning) enter the soil in solution, the calcium will tend to replace hydrogen on clay surfaces. Later, as the calcium concentration in the soil solution diminishes, and the hydrogen concentration increases, hydrogen will begin to replace calcium.

Clay mineral surfaces also can adsorb nutrient anions such as phosphate and nitrate (Wiklander, 1964). During the process of soil leaching, for example by carbonic acid, hydrogen ions replace nutrient cations on the soil surfaces, while bicarbonate anions can replace nutrient anions.

Nutrients also enter the soil bound in organic matter when leaves, fruit, and other plant parts are shed, or when an entire tree dies and falls. Dead plants and animals on the forest floor constitute the 'litter' in Fig. IV.1. This litter is equivalent to the 'dead organic matter' in Fig. I.2, and nutrients entering the soil bound in organic matter will follow the same pathways shown in the below-ground community of Fig. I.2.

Nutrients leached out of the community of soil microorganisms and humus near the soil surface are exchanged on clay, if present, in the deeper soil. If the forest is undisturbed, a large proportion of the nutrients held on clay surfaces in the mineral soil can be taken up by plant roots. These nutrients are incorporated into wood, leaves, and other plant parts.

Herbivores remove part of the nutrient stock in the leaves, fruit, and, in some cases, even stems and roots of plants. Some of these nutrients are returned to the soil in feces and urine, some are returned when the herbivores die, and some are passed up through the food chain to top predators. Eventually, all nutrients passing through the consumer food chain are returned to the soil (Lugo et al., 1974).

A proportion of the nutrients exchanged on soil colloidal surfaces is not taken up but, rather, is leached out of the soil horizon by percolating water. Water movement may be more or less vertical towards a ground water table, or lateral above an impermeable layer towards a nearby stream. In undisturbed ecosystems, the proportion lost by leaching is generally low (Finn, 1978). Most of the nutrients held in the soil are recycled by the vegetation.

However, when a forest is cut the active root biomass and, consequently, the capacity of the vegetation to recycle nutrients are reduced. If the disturbance is short-lived, newly establishing vegetation may take up some nutrients in the soil before they are leached. However, if the disturbance is long-term and vegetation regrowth suppressed, nutrient loss through leaching can be significant (Likens et al., 1970).

B. Differences between non-volatile cycles

All of the non-volatile elements have cycles qualitatively similar to that shown in Fig. IV.1. Included in this group are: the alkali metals, such as sodium and potassium; the alkaline earth metals, such as calcium; other trace elements, such as copper and zinc; and aluminum and manganese, which may be essential at low levels but which are toxic at high concentrations.

Although the non-volatile elements all have similar ecosystem cycles, quantitatively the cycles are very different. In general, monovalent nutrient cations, such as sodium and potassium, move more quickly than divalent cations, such as calcium and magnesium, which in turn move more quickly than polyvalent cations. For example, the transfer of monovalent cations from foliage to soil is primarily via throughfall, since these elements are readily leached from the leaves. In contrast, divalent cations are transferred primarily by leaf fall because they are bound more strongly. In a Puerto Rican rain forest, throughfall carried 98 percent of the potassium but only 21 percent of the calcium moving from canopy to forest floor (Jordan et al., 1972).

Leaching from the soil also depends in part on valence. For example, during an experimental slash and burn agricultural treatment in the Amazon Territory of Venezuela, 35 percent of the potassium was lost during the first year through leaching, but only 5 percent of the calcium was leached (Jordan, 1985).

Although monovalent cations often move rapidly in ecosystems, some non-essential monovalent cations move very slowly. For example, cesium uptake and release by a tropical rain forest was much slower than that of potassium and calcium (Jordan et al., 1972).

The cycle of sodium in ecosystems is similar to that of potassium, but it is seldom measured on an ecosystem scale because it rarely appears to limit plant growth. It probably is more important for animal metabolism (Frieden, 1972). Sodium levels could be critical for rain forest animals because its high mobility can result in rapid leaching (Wiklander, 1964).

In some tropical rain forest areas, there are high amounts of iron and aluminum in the soil (Van Wambeke, 1978). Naturally occurring tree species are tolerant of the high aluminum levels typical of acid tropical soils. In contrast, many agricultural crops are not aluminum tolerant. Soluble manganese can likewise be a problem for tropical crops (Baker, 1976). In contrast to iron and aluminum, other nutrient elements, such as zinc, copper, boron, and molybdenum are scarce and can be limiting under continuous cultivation (Sanchez et al., 1983).

The standing stocks and fluxes of nutrients in forest ecosystems are highly variable. For example, total calcium in the soil of an oxisol in the Amazon Basin was 7 kg/ha, while in a moist forest in Panama, 22 166 kg/ha were reported (Table III.6). In the same two ecosystems, total potassium values in the soil plus vegetation were 378 kg/ha and 3454 kg/ha, respectively (Golley et al., 1975; Jordan, 1985). The dramatic difference in the size of the total stocks suggest that the two sites have very different potentials for agricultural exploitation. Yet the potential for sustained exploitation is limited even in the Panamanian site because of the relatively large proportion (89 percent) of total ecosystem potassium stored in the vegetation (Golley et al., 1975).

Not only are nutrient stocks much greater in the Panamanian forest, fluxes due to throughfall and leaf litter fall are also much higher (Table IV.1, rows 1 and 2). Given the greater quantities of nutrients, one might expect that the proportion of the stock that is cycled would also be much greater at the Panamanian site. However, the throughfall and leaf fall data indicate that the percentages of nutrient stocks in leaves, living biomass, and in the total ecosystem that move annually as throughfall and litterfall (Table IV.1, rows 3–5) are very similar in both ecosystems, except in the case of foliar potassium.

C. Phosphorus

In the above-ground portion of the ecosystem, the cycle of phosphorus is basically similar to that of the other non-volatile nutrients (Fig. IV.1). Because of its importance in metabolic reactions, it is more highly concentrated in leaves than in wood (Fig. II.3). Consequently, seasonal loss of leaves can significantly reduce the quantity of phosphorus in a tree. However, phosphorus is relatively mobile in plants, and a proportion of the phosphorus in leaves is

Table IV.1

Fluxes of potassium and calcium in a eutrophic ecosystem (Panama) and an oligotrophic ecosystem (Venezuela) and fluxes as a percentage of standing stocks

	Moist forest,* Panama		Rain forest,† Venezuela	
	K	Ca	K	Ca
1. Nutrients in leaf litter (kg/ha^{-1} a^{-1})	129	240	6	8
2. Nutrients in throughfall (kg/ha^{-1} a^{-1})	50	37	17	4
3. 1 + 2 as a percentage of stock in leaves (%)	132	125	66	112
4. 1 + 2 as a percentage of stock in living biomass (%)	6	14	8	5
5. 1 + 2 as a percentage of stock in ecosystems (%)	5	1	5	4

* Data from Golley *et al.*, 1975.
† Data from Jordan, 1985.

often translocated into the stem before leaf abscission in many tropical trees (Herrera *et al.*, 1978b).

The rate of phosphorus input from the atmosphere is very low, compared to other major nutrient elements, and, for this reason, soil phosphorus takes on special importance (Walker and Syers, 1976).

A general classification system of phosphorus fractions in the soil is (after Williams *et al.*, 1967): (1) acid extractable phosphorus; (2) non-occluded phosphorus; (3) occluded phosphorus; (4) organic phosphorus.

Acid-extractable phosphorus is readily soluble, and is often taken to represent that phosphorus which is available to crop plants (Olsen and Dean, 1965). Non-occluded phosphorus is believed to represent phosphate ions sorbed at the surfaces of iron and aluminum oxides, hydrous oxides, and calcium carbonate. Occluded phosphorus refers to those phosphate ions present within the clay structure (Evans and Syers, 1971), or incorporated within coatings and concretions of iron and aluminum oxides during soil development. Organic phosphorus includes phosphorus held by the soil humus in labile, resistant, and aggregate protected forms (Coleman *et al.*, 1983).

During the weathering of soil minerals, important changes take place in the relative amounts of the different phosphorus fractions. In many weathering sequences, phosphorus is held largely in an acid extractable form during the early stages of soil formation. As weathering proceeds, an increasingly large proportion is held in the occluded fraction (Walker and Syers, 1976) and phosphorus becomes less readily available to plants. Because weathering processes are more intense in the tropics, phosphorus immobilization seems to be more extreme in tropical soils than in non-tropical soils derived from similar parent materials.

Many factors contribute to the fixation of phosphate in tropical soils, including: time, temperature, water content, mixing, and the presence of electrolytes. One of the most important factors is soil acidity (Fox and Searle, 1978). During the course of mineral weathering, basic cations tend to be replaced by hydrogen or aluminum with a resultant decrease in soil pH (Uehara and Gillman, 1981). Soil pH governs phosphate reactions with iron, aluminum, and manganese. As soil pH drops, these metals react with soluble phosphorus to form insoluble hydroxy phosphates. In silicate minerals, such as kaolinite, hydrous oxides of iron, aluminum and manganese occur on the clay surfaces, where they readily react with phosphate to form hydroxy phosphates (Brady, 1974). Because many soils in the humid tropics are highly weathered, acid, relatively high in iron and aluminum, and contain a large proportion of silicate minerals, phosphorus fixation is often an important phenomenon.

Soil pH can also be important in regulating phosphorous fixation in alkaline soils. In calcareous soils, calcium can react with soluble phosphorus compounds rendering the phosphorus insoluble (Brady, 1974). The relationship between phosphorus solubility and soil pH is shown in Fig. IV.2. This relationship is generalized, however, and the exact values depend on the specific nature of the compounds (Lindsay and Moreno, 1960).

One other reaction deserves mention because of its importance in otherwise nutrient-rich, volcanically derived soils. In humid regions, volcanic ash weathers quickly into allophane, an amorphous aluminum–silicate mixture that rapidly forms complexes with phosphorus and immobilizes it (Sanchez, 1976).

Phosphorus is also retained in soil organic matter, and the amount of phosphorus in tropical soils is often proportional to the amount of organic matter (Brams, 1973; Frangi and Lugo, 1985; Borie and Zunino, 1983). Phosphate can be adsorbed on the surface of soil organic matter, in the same manner as other anions (López-Hernández and Burnham, 1974). It is also

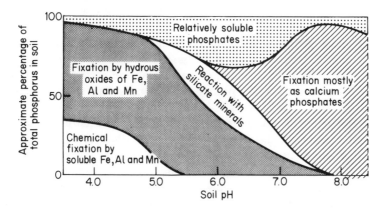

Fig. IV.2. Inorganic fixation of phosphates at various soil pH values. Average conditions are postulated, and it is not to be inferred that any particular soil would have exactly the same distribution. [Adapted from Brady (1974) with permission of Macmillan Publishing Co.]

bound in complex organic compounds within the soil organic matter itself, and often only a small proportion of this phosphorus is available to plants (Coleman *et al.*, 1983).

Despite the high proportion of phosphorus present in insoluble forms in most rain forest soils, phosphorus uptake is sufficient to maintain productivity at levels as high as or higher than those of other world forests (Fig. I.1). Certain soil microorganisms which produce organic acids that chelate iron, aluminum and manganese may play a key role in phosphorus availability (Stevenson, 1964; Graustein *et al.*, 1977). Phosphorus mobilization from recalcitrant organic matter in the soil is facilitated by oxalic acid produced by mycorrhizae (Sollins *et al.*, 1981), or the mycorrhizae can increase phosphorus uptake simply by increasing root surface area.

The mechanisms for mobilizing phosphorus all depend on living organisms in the undisturbed forest. Because of the mechanisms, the productivity of native forests is seldom seriously inhibited by lack of phosphorus. However, when the forest is cut and burned, these mechanisms of phosphorus mobilization are destroyed, as are the nutrient conserving mechanisms described in Chapter II. Consequently, phosphorus scarcity quickly becomes a problem in tropical agriculture (Olson and Engelstad, 1972).

Agriculture is possible for a few years after clearing, due to the mobilization of phosphorus in the organic matter and ash from the original forest. When the forest is burned, some of the calcium, potassium, and magnesium in stems and leaves is converted to ash and deposited on the soil surface. These nutrients are immediately available for plant uptake but, perhaps more importantly, they raise the pH of the surface soil (Nye and Greenland, 1960). One effect of increased soil pH is a decrease in aluminum and manganese toxicity (Baker, 1976). Another effect is to increase the availability of phosphorus (Fig. IV.2). The soil becomes more suitable for crop production. The effects of the ash input last from a few months to a year, depending on the amounts of ash and rainfall.

Actually, during most slash and burn operations, only a small proportion of the dried slash is converted to ash, sometimes only the leaves and small branches. The decomposing, unburned slash is also important for phosphorus supply to crops. As long as the products of organic matter decomposition are carried down into the soil, phosphate may remain relatively available (Dalton *et al.*, 1952). Eventually, however, as surface organic matter decomposes, and as basic cations in the soil are leached away, soil pH decreases, phosphorus availability declines, and crop productivity drops (Olson and Engelstad, 1972).

At low levels of available soil phosphate, weedy secondary successional species invade the cropland and out-compete crop plants, possibly because of better-adapted mycorrhizal associations or because the weed species are characterized by slower, more efficient root uptake kinetics (Chapin, 1980). A fallow period (that is, growth of successional vegetation) must occur before enough phosphorus is incorporated into the forest biomass and soil to supply the nutrient needs of another agricultural cycle (Jaiyebo and Moore, 1964).

D. Volatile element cycles

Volatile element cycles differ from those of non-volatile elements in that the atmosphere constitutes an important source and sink. For a major portion of their cycles, volatile elements are in a gaseous phase.

1. Nitrogen cycle

(a) Nitrogen fixation

The primary input of nitrogen into ecosystems is through the processes of biological nitrogen fixation. In this process, atmospheric nitrogen is reduced to ammonium (Delwiche, 1977).

Nitrogen fixation is carried out by a wide variety of symbiotic and free-living organisms (Stewart, 1977). The most familiar association, and the most important in temperate croplands, is that between root nodulated legumes and nitrogen-fixing bacteria of the genus *Rhizobium* (Stewart, 1967). Nitrogen fixation by nodulated trees of tropical rain forests can contribute nitrogen to the forest, but often fixation by other organisms is quantitatively more important (Mague, 1977; Stewart *et al.*, 1978).

A second type of association involves non-leguminous root nodulated plants. In the temperate zone, the most common association is between angiosperms, such as alder, and actinomycete bacteria. In the tropics, there is an association between primitive gymnosperms, such as the cycads, and nitrogen-fixing blue-green algae (Stewart, 1967).

Nitrogen-fixing blue-green algae also form symbiotic relationships with certain fungi, liverworts, and ferns. The algal–fungal associations are called lichens. Such associations are common in tropical rain forests (Forman, 1975). Free-living and symbiotic nitrogen-fixing bacteria also are common on leaves and in the soil of tropical rain forests (Stewart, 1977). They are even found in the digestive tract of termites (Benemann, 1973). Measurements of the rates of nitrogen fixation by microorganisms on leaves and in the soil indicate that their contribution to the nitrogen budget of tropical rain forests is considerable. Annual nitrogen fixation rates can be greater in the wet tropics than in other regions (Table IV.2) because of the year-round warm and wet climate.

Two other nitrogen-fixing associations are worth mentioning here because of their economic importance in the tropics. One is the association between the water fern, *Azolla*, and blue-green algae in flooded rice fields of the Far East. When the water level subsides and the soils dry up, the *Azolla* dies and its nitrogen is released for use by the rice (Soermarwoto, 1977). The second is between several tropical grasses and at least two genera of nitrogen-fixing bacteria. Exploitation of this association holds promise for decreasing problems of nitrogen fertilization in the tropics (Döbereiner, 1977).

Nitrogen can also enter ecosystems through dry-fall or dissolved in

Table IV.2

Nitrogen cycling in the tierra firme forest compared to other tropical and temperate forests: pools are given in kilograms of nitrogen per hectare and fluxes are given in kilograms of nitrogen per hectare per annum; brackets indicate summed compartments (from Jordan *et al.*, 1982)

	Amazon rain forest, Tierra firme	Amazon rain forest, Caatinga on podsol	Seasonal forest, Banco I, Ivory Coast	Hardwood forest, Coweeta, North Carolina	Hardwood forest, Hubbard Brook, New Hampshire	Douglas Fir, Andrew Forest, Oregon
Pools						
Leaves	143	}336	}1150	}995	}351	144
Stems, branches and bark	941	834	—	—	181	394
Roots	586	132	—	140	1100	197
Litter and superficial humus	406					798
Soil	3507	785	6500	6917	3626	3397
Total	5583	2087	7650	8052	5258	4930
Inputs						
NH₄–N in precipitation	11.3	}21.0	}21.2	2.7	}6.5	}2.0
NO₃–N in precipitation	.2			3.6		
N–fixation	16.2	>35	—	12.0	14.2	2.8
Outputs						
NH₄–N leached	8.4	}9	}21.2	0.06	}4.0	}1.5
NO₃–N leached	5.7			0.1		
Denitrification	2.9	—	—	10–18	—	—
Balance (input – output)	+8.9	—	—	~0	+16.7	+4.3
Internal fluxes						
N in leaf-fall	61.3	24	170	33	54.2	10.8
NH₄–N in throughfall	25.0	}9	}80	}4	}9.3	}3.4
NO₃–N in throughfall	0.3					
Total fluxes	86.6	33	250	37	63.5	14.2

Note on pool groupings: for Coweeta the bracket }995 sums leaves + stems, with roots — and litter 140; for Hubbard Brook the bracket }351 sums leaves + stems, with roots 181 and litter 1100; for Banco the bracket }1150 sums leaves through litter.

precipitation, just as other nutrients do (Visser, 1964; Söderlund, 1981). This nitrogen may be incorporated in organic compounds or it can be in the form of nitrate or ammonium. Ordinarily, nitrogen in dry-fall and precipitation is relatively unimportant. However, following the burning of tropical forests, ammonium and nitrate input by way of wet and dry fall can be very high (Lewis, 1981). The source of the nitrogen is probably biomass volatilized during the burning. Nitrogen deposited from the atmosphere may also originate by lightning fixation during thunderstorms (Drapcho *et al.*, 1983).

(b) Ecosystem cycle

Nitrogen which has entered from the atmosphere begins a cycle within the ecosystem (Fig. IV.3). The nitrogen is incorporated into amino acids and then into plant proteins. Some of the nitrogen then passes through the food chain (Fig. IV.1), and it is incorporated into new proteins. When plants and animals die, organic matter returns to the soil where decomposer organisms break down complex nitrogenous compounds into simpler compounds (Delwiche, 1970).

The rate at which complex compounds are broken down depends, in part, on temperature and moisture availability (Singh and Gupta, 1977). Since temperature and moisture are near optimum year-round in the humid tropics, organic matter breakdown can be relatively fast. Breakdown also depends on the ratio of carbon to nitrogen in the litter itself. When the carbon:nitrogen ratio is relatively high, as in the wood of some of recently fallen trees, decomposition is slow because the concentration of nitrogen is insufficient to supply the

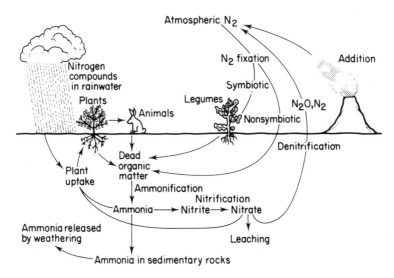

Fig. IV.3. A simplified representation of the nitrogen cycle in a terrestrial ecosystem. [Adapted from Jackson and Raw (1973) with permission of Edward Arnold Publishers.]

metabolic needs of many decomposers. The decomposition of such materials is aided, in part, by fungi whose hyphae can spread to other more nutrient-rich sources, such as decomposing bark. Eventually, as carbon is respired or leached away, the carbon:nitrogen ratio declines to the range between 20:1 and 10:1. Nitrogen is then less limiting to decomposers, and decomposition proceeds more rapidly (Swift et al., 1979).

In tropical rain forests, termites are often important agents in the breakdown of organic matter, especially that of large trees (Matsumoto, 1976). Although some termite species actually increase the carbon:nitrogen ratio of the material they process, they also increase the surface area of the organic matter, and they break down complex molecules, such as lignin, thereby increasing the opportunity for the activity of other decomposers. Other termite species, such as those which cultivate fungal gardens, decrease the carbon:nitrogen ratio of the material they process (Wood and Sands, 1978).

One of the last transformations in the sequence of organic matter decomposition is the conversion of the nitrogen in amino acids to ammonium (Runge, 1983). The ammonium ion has a positive charge, and can either be taken up by plant roots or can be exchanged on the surface of humus or clay particles, as are other cations. Ammonium can volatilize directly into the atmosphere, leach downward through the soil, or undergo nitrification (Bolin and Arrhenius, 1977).

Nitrification is the oxidation of ammonium to nitrite and then nitrate through the activity of nitrifying bacteria. Nitrification sometimes appears to be inhibited in mature ecosystems, both in the temperate zone (Rice and Pancholy, 1972) and in the tropics (Jordan et al., 1979a). Inhibition may be a nitrogen conserving mechanism (Vitousek et al., 1979), but the mechanism by which nitrifiers are controlled in mature ecosystems is not known. Allelochemical inhibition has been proposed, but the evidence is still fragmentary and disputed (Verstraete, 1981; Lamb, 1980).

When ecosystem disturbances are large enough to modify microclimates and destroy the organic matter which may inhibit nitrification, the rate of conversion of ammonium to nitrate increases. Nitrate can be taken up by plants or held on anion exchange sites. Nitrate can also be leached downward and lost from the ecosystem to drainage streams, or it can be converted to gaseous, molecular nitrogen through the activity of denitrifying bacteria (Delwiche, 1977).

(c) Regional differences in nitrogen stocks

While soil stocks of calcium and potassium are usually lower in lowland tropical forest ecosystems than in temperate broad-leaved forest ecosystems, there is less difference in soil nitrogen stocks (Fig. II. 4(a), (b)). This is due to differences in the way the two types of nutrients are stored. Calcium and Potassium are exchanged on the surfaces of humus and clay colloids where they can be replaced by hydrogen or aluminium ions. In contrast, nitrogen is stored primarily in undecomposed organic matter (Brady, 1974), where it is less readily leached. Nevertheless, there is some difference caused by lower rates of decomposition in cooler regions. Annual fluxes into and out of the soil nitrogen pool probably are much greater in tropical forests than in temperate forests (Table IV.2; Greenland, 1958). This is because the year-round high temperatures and high humidity of the humid tropics permit continual microbial activity and, consequently, continual nitrogen fixation, volatilization, and denitrification. The rapid and continuous conversion of organically bound nitrogen to soluble forms may account for Vitousek's (1984) data that suggest nitrogen is seldom limiting in undisturbed lowland humid tropical forests.

In contrast, nitrogen frequently seems to be limiting in regions with a pronounced cold season or with a Mediterranean climate (Vitousek, 1982). In regions with a Mediterranean climate, drought during the summer inhibits decomposition and nitrogen mineralization (Jenny, 1950), and at high latitudes and altitudes, low temperatures inhibit breakdown of litter. Nitrogen limitation in cold or dry regions may result from the lag between growth onset in higher plants and the nitrogen availability, because tree roots can respond more quickly than microbial populations to favorable conditions (Post *et al.*, 1982; Zinke *et al.*, 1984).

(d)   Changes in nitrogen fluxes due to disturbance

Rapid breakdown rates of nitrogen compounds in the soil of undisturbed tropical forests do not lower productivity because high rates of nitrogen loss are compensated by high rates of nitrogen fixation. However, forest clearing has a major impact on nitrogen stocks and dynamics. This is so because the processes which contribute to high rates of nitrogen loss are unabated or even increase, while nitrogen accretion processes are diminished or even completely stopped.

Cutting and burning of tropical forests eliminate the whole living complex of trees and microorganisms which fixes and conserves nitrogen. Ammonium, which in the undisturbed forest is rapidly taken up by trees, is quickly converted to nitrate, since nitrifying bacteria no longer compete with tree roots. Crop species, if present, take up only a small proportion of the nitrate which becomes available. The rest of the nitrate is lost either through nitrate leaching or through denitrification (Nye and Greenland, 1960). Detailed studies showing the increase of nitrification and nitrate leaching following deforestation in a northern hardwoods ecosystem have been presented by Smith *et al.* (1968), Bormann *et al.* (1968), and Likens *et al.* (1969).

2. Sulfur cycle

The sulfur cycle in terrestrial ecosystems is similar to that of nitrogen, in that the atmosphere is a principal source and sink. Sulfur in the atmosphere occurs principally as sulfur dioxide (SO_2) and aerosols of sulfate (XSO_4) (Kellog *et al.*, 1972). Sulfates can be carried into an ecosystem dissolved or suspended in precipitation, or may settle out as dry-fall (Galbally *et al.*, 1982). Sulfate, once it reaches the soil, can be taken up by plant roots.

Sulfur dioxide, like sulfate, can be carried into ecosystems dissolved in rainwater. Direct uptake of sulfur dioxide through the leaves of some species can be quantitatively important (Galbally *et al.*, 1979).

Sulfur is a key constituent of several amino acids and also plays a role in plant and animal metabolism (Bowen, 1979). Following the death of plants and animals, the complex sulfur-containing organic compounds are gradually oxidized to sulfate by decomposer bacteria (Ivanov, 1981). The sulfate is then either recycled by plants and decomposers, or is leached down into the mineral soil where it can be held on the anion exchange sites of certain clay minerals. There it can be stored, lost through leaching, or taken up and recycled in the ecosystem (Johnson *et al.*, 1982a).

Sulfur input into undisturbed ecosystems has been observed to be greater than leaching losses in both temperate (Johnson *et al.*, 1982a) and tropical (Johnson *et al.*, 1979) forests. The excess sulfur may be transformed into organic forms by microbial populations (Swank *et al.*, 1984), adsorbed as sulfate on clay surfaces (Johnson *et al.*, 1982a), or possibly volatilized as H_2S (Delmas *et al.*, 1980).

Although there is no evidence that sulfur is limiting in undisturbed tropical forests, losses of sulfur can be very important during forest conversion. In a Costa Rican forest which was cut and burned for agriculture, 31 percent of the total sulfur in the ecosystem was removed by selective harvest, 44 percent was volatilized during burning, and 6 percent was lost by leaching (Ewel *et al.*, 1981). Sulfur has the potential to limit the production of tropical crops.

E. Chapter summary

This chapter examined the characteristics of the cycles of individual nutrient elements in tropical rain forest ecosystems. Those characteristics which contribute to the rapid loss or immobilization of nutrients following forest disturbance were emphasized, as a background for the material in the following chapter.

An important aspect of the potassium cycle is this element's relative mobility in the ecosystem, especially during forest clearing. Cutting of the forest destroys nutrient conserving mechanisms and can result in rapid leaching losses of potassium.

The role of calcium, magnesium, and potassium in regulating soil pH and, thus, the availability of phosphorus, is important in the context of slash and burn agriculture. Following cutting and burning, the levels of these cations are high in the soil and, thus, these nutrients, as well as phosphorus, are available for crop uptake. However, calcium, magnesium, and potassium are rapidly leached because of heavy rain and year-round microbial activity. These elements soon become scarce, soil pH declines, and phosphorus availability decreases.

Aluminum and manganese, which, at high concentrations, are toxic to crop plants, are also affected by soil pH, and thus by the cycles of calcium and potassium. When soil pH is low, aluminum and manganese are soluble and readily taken up by plants. As pH increases, their solubility decreases.

Deforestation strongly affects the nitrogen budget. Nitrogen cycles rapidly in the wet tropics, and when the forest is cut, processes which contribute to nitrogen loss continue unabated, while nitrogen fixation decreases. Deforestation also causes high sulfur losses.

An understanding of these control mechanisms provides a basis on which to compare the effects of different types of disturbances on nutrient cycling in tropical rain forest ecosystems.

Chapter V

Changes in nutrient cycles due to disturbance

The first chapter in this book developed the argument that the potential for nutrient loss in wet or moist tropical rain forests is relatively high due to the continuous hot and humid climate. Then it was shown that, despite the high potential for nutrient loss in wet tropical forests, actual loss is usually quite low, and does not pose a threat to continuous ecosystem functioning, due to the nutrient conserving mechanisms of the undisturbed native forests. The third chapter showed that these mechanisms are most highly developed on those soils extremely low in one or more critical nutrient elements.

These discussions concerned nutrient cycling in naturally occurring tropical forests. In this chapter, case studies are presented which illustrate changes in nutrient dynamics following natural or man-caused perturbations which disturb or destroy the nutrient conserving mechanisms. These examples will help answer questions about tropical forest ecosystems, such as: What is the effect of disturbance on the productive potential of tropical ecosystems? How long can productivity be sustained in disturbed ecosystems? What effect does disturbance type have upon ecosystem response? How does disturbance affect the ability of an ecosystem to return to pre-disturbance condition?

A. What is disturbance?

Before beginning a discussion about ecosystem response to disturbance, it is important to know what a disturbance is. The answer to this question is neither simple nor obvious. In fact, distinguishing between disturbed and undisturbed forests may not be possible. To explain, we can begin by considering forests thought to be 'undisturbed'.

An 'undisturbed' forest is sometimes taken to mean a forest that is in 'steady state', that is, one in which biomass, number of trees per hectare, and species composition are unchanging. Further, the rate of nutrient input into such a system is balanced by the nutrient output. Such systems have also been called 'climax' ecosystems. However, the terms 'steady state' and 'climax' are not really appropriate, because ecosystems are always changing, even when completely unaffected by man or natural disaster. Ecosystem changes occur at many scales, with cycles ranging from diurnal through seasonal and yearly, to longer term cycles that span decades or centuries.

Diurnal changes include those in the metabolic rates of organisms in response to daily cycles of temperature and light. Photosynthesis, of course, closely follows the daily cycle of light and also the yearly cycle of temperature. Many animals, for example some insects, are more active during the day because their metabolism responds to heat, but others move around at night when there is less probability of capture by predators.

Seasonal changes, such as pulses of leaf fall, flower, and fruit formation and animal migration, are adaptations to changes in temperature and day length that are caused by the Earth's annual rotation around the Sun. A forest that is exposed to seasonal changes but which maintains constant structure and function from one year to the next has been termed 'climax' (Weaver and Clements, 1929). However, we now know that there really are no unchanging forests. Unusual events such as volcanic eruptions and dust storms produce atmospheric aerosols which are capable of modifying both energy and nutrient inputs to ecosystems around the world (Toon and Pollack, 1980). Decrease in energy input results from aerosols blocking solar radiation. For example, the volcanic eruption of Mt Tambora, Indonesia, in 1815 produced such large quantities of ash and dust that light reaching the Earth was reduced sufficiently to lower surface temperatures for several years (Stommel and Stommel, 1979; Stothers, 1984). An increase in nutrient input into ecosystems results from dust storms. Such storms in the southwestern United States have caused sporadic inputs of phosphorus to the deciduous forests of the eastern United States (Swank, 1984). Dust from the Sahara is frequently deposited in South America (Prospero et al., 1981).

Factors within ecosystems also can cause cycles. Periodic widespread insect infestations, such as the spruce-budworm in the taiga forests of North America, may have a large impact on ecosystem structure and function (Mattson and Addy, 1975). Swank et al. (1981) have reported an increase in nitrate export following insect infestations.

Another example of the difficulty in identifying 'undisturbed' forests can be seen in some lowland regions of Central America. Until recently the area was thought to be primeval jungle, but there is now ample evidence that 1000 years ago it was occupied by the highly developed Mayan civilization and that present-day forests occupy former croplands (Hammond, 1982).

One way of ascertaining whether an ecosystem has been recently disturbed is to look at certain characteristics of the plant community. Recently disturbed forests are typically dominated by fast growing tree species with low wood density, low resistance to insects and disease, and a relatively short life-span. Seeds of these species are often small and light and are dispersed by the wind, or by birds and bats. This type of forest is usually called 'secondary successional' forest.

In contrast, forests that have not been severely disturbed for a long time are dominated by trees whose wood volume increases slowly. The specific gravity of the wood is very high. They are more resistant to insects and are longer lived. The seeds of many such species are heavy and are often dispersed by mammals. Forests of this type are called 'primary' forests.

Time since disturbance is not the only factor to consider in classifying forest as disturbed or undisturbed. There is also a problem of scale. For example, a hurricane that knocks down hundreds of hectares of trees certainly is a disturbance, but is the opening in a forest caused by the fall of a single tree a disturbance? Since such openings, or 'gaps', are sometimes colonized by successional species, gaps perhaps should be classified as disturbances. Yet gaps occur in all forests. It may not be meaningful to talk about 'undisturbed' ecosystems, since all ecosystems are disturbed, if only by gap formation.

Although all ecosystems are disturbed to a lesser or greater extent, disturbances vary greatly. Ecosystem disturbances can be classified according to: (1) intensity; (2) size; (3) duration.

B. Classification of disturbances

1. Intensity

The intensity of disturbance can be light, moderate, or severe. A light disturbance is defined, here, as one which does not disrupt the basic structure of the ecosystem. A tree-fall gap created when a large tree dies and falls, is an example of a light disturbance. Another example is careful selective logging where only a few trees per hectare are cut and removed without the construction of roadways.

A moderate disturbance can be defined as one in which the structure of the forest is destroyed, but the soil is not degraded. An example is the cutting of primary forest and the planting of tree crops. Moderate disturbances occur in the tropics when native forests are replaced by plantations of coffee, cacao, rubber, or pines.

A severe disturbance is defined as one in which forest structure is destroyed and the soil is severely degraded. Both lava flow from volcanoes and forest clearing with heavy machinery remove or bury topsoil, and thus are severe disturbances. Overgrazed pasture, where the activity of animals results in loss of topsoil through erosion also can be classified as a severe disturbance.

2. Size

The factor used here to classify disturbance size is the ease with which seeds that initiate the recovery process can enter the disturbed site. In small disturbances such as tree-fall gaps, seeds from surrounding trees can fall directly into the gap, or were already in the soil before gap formation.

In intermediate sized disturbances, seed-dispersing animals freely traverse the disturbed area. Most abandoned slash and burn agricultural sites in the tropics are small enough (one to a few hectares) that wildlife freely enters. Wildlife may also be important vectors of mycorrhizal spores (Mosse *et al.*, 1981).

In large disturbances, the distance from undisturbed forest to the middle of the disturbed area is beyond the normal range of the animals which carry seeds and mycorrhizal spores. Revegetation may be by windborne seeds (Whitmore, 1983) such as those of many grasses and herbs (Uhl and Clark, 1983).

3. Duration

A short disturbance is a single, discrete occurrence such as a hurricane or a logging operation, which is over in a matter of days, or less, and after which recovery begins immediately.

An intermediate duration disturbance can be a series of disturbance events, such as planting, weeding, and harvesting during shifting cultivation. When the disturbance ceases, as when the agricultural plot is abandoned, recovery occurs without interruption.

A long-term disturbance is one in which disruptive effects continue long beyond the cessation of the original disturbance event. An example is over-grazed pasture. Trampling by livestock compacts the soil and prevents the infiltration of water. Consequently, re-establishment of vegetation is inhibited, erosion begins, and degradation continues long after the cattle have been removed.

C. Case studies of disturbances

1. Very light intensity: World-wide fallout

During the 1950s and early 1960s, nuclear weapons were routinely tested in the atmosphere. As a result of these tests, radioactive fallout contaminated virtually all the ecosystems in the world (Wolfe, 1963). The impact of radioactive fallout was potentially so significant that a large research program was initiated to determine possible fates and effects of radioisotopes deposited in ecosystems (Auerbach, 1971).

One of the most important findings from these studies was that radio-isotopes, after being desposited on leaves and soil, cycle through ecosystems in the same manner as stable elements (Reichle *et al.*, 1970). Many of the radioisotopes are passed up through food chains, and could be ingested by humans. In some cases, isotopes are concentrated as they move through the food chain. Cesium-137 is an example of a radioactive isotope which was found to become concentrated in this way (Lindell and Magi, 1967; Nevstrueva *et al.*, 1967). In the Arctic tundra, cesium-137 fallout was adsorped on the surface of

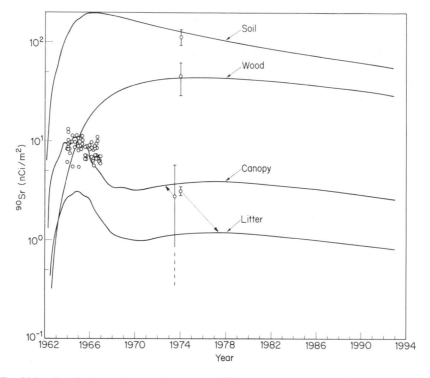

Fig. V.1. Predictions of the total amount of ^{90}Sr in various compartments of a tropical rain forest from 1962 through 1994, and validation points for 1966 and 1974. Note that the total amounts in each compartment are plotted. If the total amounts were divided by the mass of each compartment to give isotope concentrations, soil and wood would be lower than canopy leaves. [Adapted from Jordan and Kline (1976) with permission of the Health Physics Society.]

lichens, which act similarly to an ion-exchange column (Witkamp, 1966). The lichens, which form an extensive ground cover, are a principal food of the region's caribou. The cesium ingested along with the lichens was found to accumulate in muscle tissue of the caribou. The caribou, in turn, are an important food of native Arctic people, and due to food chain accumulation, body burdens of cesium-137 increased to levels which were considered near the maximum permissible (Liden and Gustafsson, 1967; Miettinen and Hasanen, 1967). These studies provided important evidence which led to the signing of the atmospheric nuclear test ban treaty in 1963 (Whicker and Schultz, 1982).

Despite the potential health significance to top predators from radioactive fallout, the levels of radioactivity which occurred as a result of worldwide fallout were never high enough to noticeably affect ecosystem structure or function. To the casual observer, and even to most scientists, worldwide radioactive fallout had no discernable effects on ecosystems. Nevertheless, radioactive fallout did affect element cycling. There was a change in the stocks and dynamics of radioactive isotopes, which are normally present in all ecosystems at background levels. A study which illustrates these changes was that of the cycling and retention of strontium-90 in a Puerto Rican rain forest (Jordan et al., 1973), Strontium is a particularly hazardous isotope because it can substitute for calcium in metabolic processes (Comar and Lengemann, 1967). The study predicted amounts of strontium-90, as a function of time, following the cessation of major atmospheric nuclear tests after 1963 (solid line, Fig. V.1). The predictions were based on a mathematical model derived from cycling studies of stable strontium. The predictions were compared to data from actual field measurements of strontium-90 in 1966 and 1974 (circles, Fig. V.1).

It is clear that the cessation of atmospheric tests in 1963 was important in reversing the trend of increasing radioactivity in the environment. Also clear is the relatively long retention time of the isotope in the ecosystem, even after the cessation of nuclear testing (environmental half-life was 20 years). This means that when pollutants with long physical half-lives such as strontium-90 or chlorinated hydrocarbon pesticides contaminate an ecosystem, they pose a long-term danger.

This type of perturbation, although it has little apparent impact on eco-system structure and function, nevertheless has potentially great impact on humans and other valued species.

2. Light intensity, small size, short duration: *tree-fall*

Tree-fall is a natural phenomenon in all forests. As trees age, their growth rates slow and they become more susceptible to disease or insect attack. Eventually the trees die. Usually they fall over, but occasionally a tree will decompose almost completely without falling. When large tropical trees fall, they may pull down several nearby trees because the canopies are tied together by lianas.

Gap formation alters nutrient cycling in that portion of the forest. Tree-fall gaps permit direct sunlight to reach the forest floor during part of the day (Whitmore, 1978) which increases rates of decomposition. The trunk and leaves of the fallen tree contain nutrient elements which enrich the forest floor in the gap site. This sudden input of nutrients to the forest floor might be expected to produce a sudden pulse of nutrient leaching from the litter layer to the mineral soil. Yet in a tropical rain forest, nutrient leaching from artificially created gaps did not increase (Uhl, 1984). The lack of increased nutrient leaching in forest gaps could be the result of nutrient uptake by (1) sprouts; (2) existing saplings; (3) microbes; or (4) new seedlings. Each of the mechanisms has maximum effectiveness at different times following tree fall.

(1) If a falling tree breaks off above ground level, the root system may survive and the stump send up sprouts. Other trees broken in the fall may also sprout. Sprouting stumps can take up nutrients efficiently, because they retain the large root biomass of the original tree (Luvall, 1984). Sprouting trees would have maximum effectiveness immediately after tree fall.

(2) In many tree-fall gaps, the opening heals through the growth of saplings which existed before gap formation (Whitmore, 1978). However, it is unlikely that the roots of such saplings are able to fully exploit the nutrient pulse, because their biomass is still small. It may take several years for saplings to increase greatly their nutrient uptake.

(3) Some nutrients probably initially go to the decomposer food chain. Decomposers might respond strongly within days or weeks to a pulse of nutrients. As the gap closes, decomposers would take a smaller proportion of available nutrients, with more going to trees.

(4) Sometimes the falling tree damages most of the saplings in its path. In this case, tree seeds already at the site or those which arrive subsequently, may germinate and grow up through the tangle of slash (Uhl and Clark, 1983). These may be seeds of either primary or secondary species. Uptake of nutrients by such recently germinated trees would be relatively low for many years.

A forest tree fall creates a variety of micro-sites, each of which recovers through different mechanisms. For example, trees which uproot when they fall expose a patch of mineral soil. These patches are often colonized by secondary successional species whose seeds are blown or carried to the spot or were already present in the soil. The area under the fallen canopy is a micro-site which favors already existing saplings. On the stem of the fallen tree itself, seeds of secondary successional species can germinate, and derive their nutrients from the decomposing trunk. Eventually, the trunk disappears, but by that time the young trees have sent their roots down into the soil. Some of the stilt-rooted trees encountered in the wet tropics may have originated in this way.

An interesting method of dating the formation of tree gaps has been used in a Mexican rain forest (M. Martínez, personal communication). A common local palm *Astrocaryum mexicanum* exhibits a high correlation between stem length and age. When a gap is formed, some of these palms may be bent over by the falling tree. Later the palms resume vertical growth. The age of the gap can be determined by measuring the distance from the tip to the sharp bend in the stem.

3. Light intensity, intermediate size, short duration: *wind storms*

Large, old trees usually do not die suddenly. Rather bacteria and fungi invade the heartwood creating a cylinder of rotten wood inside the tree. Termites also play a role in the decomposition. Eventually, the tree dies and falls over.

Sometimes a windstorm is the immediate cause of the fall of a weakened tree. The tangle of vines in the canopy often results in several trees going down at the same time and the formation of a relatively large gap. Occasionally, a windstorm can blow over a large number of trees at the same time. For example, in the upper Rio Negro region of the Amazon Basin, Uhl (1982) reported,

In February 1981, a wind funnel hit a patch of mature tierra firme forest, knocking over 49 trees greater than 20 cm diameter in 0.41 hectares. Trees fell in all directions. In single tree falls, most trees (in this area) snap at the base, but in this multiple tree fall, 80 percent of the trees were uprooted. A patch of approximately 300 m^2 was left untouched in the middle of the disturbed area.

Occasionally, a hurricane or tropical typhoon can level large areas of forest. For example, in 1932 a severe hurricane leveled large areas of forest in the Luquillo Forest in eastern Puerto Rico. A series of surveys over the next several decades showed a normal sequence of secondary succession (Crow, 1980). During the first few years, Basal area and volume in the disturbed area increased rapidly as species with relatively light wood filled in the area. Later, primary forest species became dominant, and by the late 1970s the structure and species composition of the disturbed area were approaching those of typical mature montane forests in the Caribbean region (Crow, 1980). No studies of nutrient losses were carried out following the hurricane, but apparently the early colonizers and surviving small trees were able to take up and store sufficient nutrients to permit the eventual re-establishment of primary forest.

On August 29, 1979, Hurricane 'David' hit the island of Dominica in the Caribbean. The damage was the most extensive ever reported (Lugo *et al.*, 1983). Forty-two percent of the standing timber was severely damaged. Nevertheless, regrowth of a large number of species was rapid. Cyclic plant successions appear to be a common phenomenon in this area where hurricances are frequent (Lugo *et al.*, 1983).

4. Light intensity, intermediate size, long duration: *shelterwood forestry*

During the first half of the 19th century, management systems for the naturally occurring tropical forests of Africa and South-East Asia were developed (Baur, 1964). In Africa, the systems were often referred to as the 'tropical shelterwood system', and in South-East Asia as the 'Malayan uniform system' (Fox, 1976). These systems were designed to promote the establishment, survival, and growth of the seedlings and saplings of desirable species by the poisoning of undesirable trees and the removal of vines and weeds. After several years, when surveys showed that the reproduction of desirable species was well established, the canopy trees of the timber species were harvested. The tract was then monitored, and weedings and thinnings took place as needed. The system was abandoned in the 1960s, partly because it did not make sufficiently intensive use of the land to compete with other forms of land use such as cocoa, oil palm, or agricultural crops (Lowe, 1977).

Although these systems appeared impractical in the face of the intensive pressure from expanding human populations, they were very desirable from an ecological point of view. Management systems which encourage natural regeneration usually are not seriously affected by disease and soil fertility because the naturally occurring species are adapted to existing conditions. A tree nursery does not have to be maintained as it does with plantation forests. Erosion usually is not a problem, since bare soil is seldom exposed on a large

scale. The natural forest provides good cover and food for game animals
hunted by the native population. In addition, the system may result in genetic
improvement of the desirable species. From the aspect of nutrient cycling, the
systems are desirable, because they do not seriously interfere with the naturally
occurring nutrient recycling mechanisms (Poore, 1976; Synott and Kemp,
1976; Jordan and Herrera, 1981; Jordan and Farnworth, 1982).

Apparently no nutrient cycling studies were made in forests under the
shelterwood or uniform systems of management. However, a recently com-
pleted study in lower montane rain forest in Costa Rica (Parker, 1985) showed
that soil water nutrient concentration below cut areas was proportional to the
size of the cut (Fig. V.2). In small cuts, such as those of shelterwood forestry,
roots from the trees surrounding the cut apparently are able to utilize some of
the nutrients before they are lost through leaching.

There is another forestry system, which does not seriously disrupt the
nutrient cycle of the forest, but is nevertheless usually undesirable. It is
selective cutting, in which the most desirable individuals of the economically
valuable species are harvested. The result is to encourage the reproduction of
undesirable species and of inferior individuals of valuble species. One way to
avoid these effects is to plant trees of the best genetic stock in the gaps created
by selective cutting (Wadsworth, 1981).

5. Moderate intensity, small size, short duration

A moderate intensity disturbance differs from one of light intensity in amount
of disturbance to basic forest structure. The shelterwood logging system and

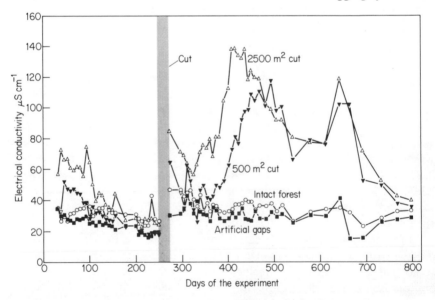

Fig. V.2. Electrical conductivity, an index of ionic concentration in soil water, in an
undisturbed control forest, single tree gaps, and cuts of 500 and 2500 m², before and
after cutting, in a pre-montane rain forest of Costa Rica. Data are from Parker (1985).

hurricanes were classified as light disturbances. Although these disturbances remove some canopy trees, saplings and small trees are left intact and, as a result, nutrient retention and recycling mechanisms remain. In this section moderate intensity disturbances, which destroyed forest structure, are described.

(a) Radiation experiment

Experimental irradiation of a rain forest was undertaken in 1965 to determine the effects of a predetermined and controlled amount of radiation on a tropical ecosystem (Odum, 1970b). The experiment was sponsored by the US Atomic Energy Commission as part of a program to predict the consequences of accidental or purposeful release of radiation into the environment. Leakage of radioactivity from a nuclear power plant is one example of an accidental release. A purposeful release could come about through the use of a thermonuclear device to excavate a new Central American canal, an idea that was under consideration at one time but later abandoned (Martin, 1969; Kaye and Ball, 1969).

In the experiment, a 10 000 Ci source of radioactive cesium was used to irradiate a lower montane rain forest in Puerto Rico for three months (Odum

Fig. V.3. Appearance, in 1966, of a Puerto Rican rain forest that was irradiated in 1965. The tripod of pipes was used to support the radiation source.

98

Fig. V.4. Appearance, in 1982, of the same plot as shown in Fig. V.3. The worker is standing under the tripod.

and Drewry, 1970). Cesium-137 emits gamma radiation which resembles solar radiation, in that shielding the source after the experiment eliminates radiation in the environment. Irradiation was carried out from January through April 1965. The forest was killed for a radius of about 15 m from the radiation source (Fig. V.3). Immediately afterwards, nutrient concentrations in the soil increased due to the pulse of nutrients from leaf fall as the forest died. However, by September 1965, nutrient levels had decreased to levels below those of the undisturbed forest (Edmisten, 1970).

By 1982, tree biomass was 60 t/ha (Silander, 1984). The secondary successional forest had almost completely obscured the original opening (Fig. V.4). Thus, while a disturbance of moderate intensity, small size, and short duration may result in some nutrient leaching, the total losses in this experiment were not great enough to inhibit forest recovery.

(b) Manual clearing

Harcombe's (1977a, b) conclusions about moderate disturbance, from a study on andosols in Costa Rica, were similar to those reached in the Puerto Rican irradiation study. He manually removed all the vegetation from a series of plots and then allowed some to revegetate naturally while others were kept bare by continual weeding. Bare plots lost more nutrients than revegetating plots, but

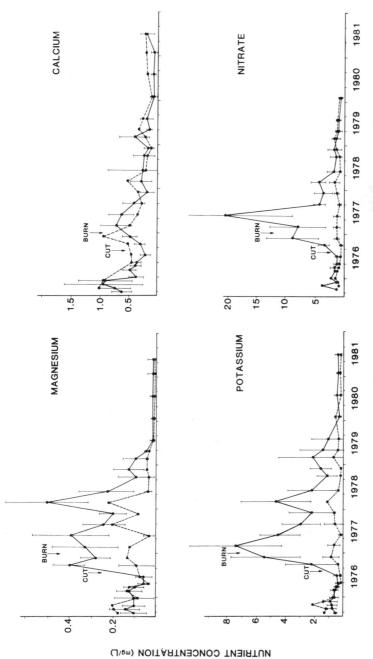

Fig. V.5 Nutrient concentrations in the soil water of a cut, burned, and abandoned plot (———) and a control plot (– – –) in the Amazon Territory of Venezuela: Confidence intervals are ± 95 percent. [Adapted from Uhl and Jordan (1984), with permission from the Ecological Society of America.]

100

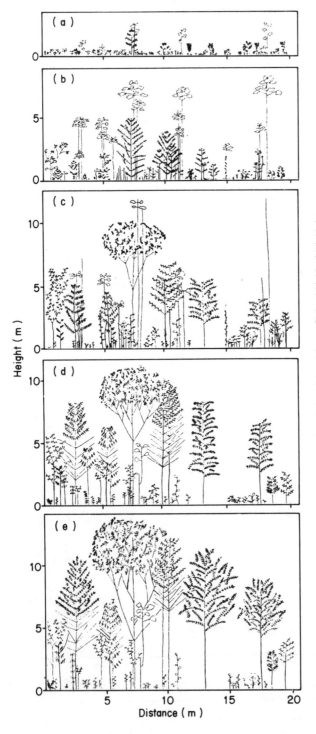

Fig. V.6. Scale drawings representing the plants present on a permanent transect during the first five years of succession following cutting and burning of a rain forest plot in the Amazon Territory of Venezuela: (a) one year; (b) two years; (c) three years; (d) four years; (e) five years. [Adapted from Uhl and Jordan (1984) with permission from the Ecological Society of America.]

even in the latter there was some initial nutrient loss. Despite nutrient losses, net biomass accumulation in the vegetated plots was very high, 1551 g m² the first year. Harcombe concluded that any nutrient loss due to manual clearing did not affect productivity.

(c)   Cut, burn and abandon

The sequence of cut, burn, and abandon is not common in tropical forests. If a forest is cut and burned, it is usually done to prepare the land for agriculture or pasture. Nevertheless, it does occasionally occur. For example, when rain forest is cleared for pasture, but for some reason grass does not establish, the area may be abandoned. The data presented here come from a study in the Amazon Territory of Venezuela, where a forest plot on oxisol was cut, burned, and then abandoned as a companion experiment to an agricultural study (Uhl and Jordan, 1984).

Following the burn, the concentration of calcium in soil water percolating out of the root zone did not increase above levels in the control plot (Fig. V.5). Likewise, phosphate phosphorus showed no significant increase in leaching losses compared to the control plot after the burn (Uhl and Jordan, 1984). The responses of magnesium, potassium, and nitrate-nitrogen were much more pronounced (Fig. V.5).

The first year after the burn, only 55 g of above-ground biomass accumulated on the site per square meter, but this poor growth probably reflected high seedling mortality due to exceptionally dry weather immediately after the burn rather than any lack of nutrients. Over 900 g of additional biomass accumulated per square meter during the second year, while living trees in the control forest accumulated biomass at the rate of about 600 g/m². During the third through fifth years, there was rapid structural development of the successional forest (Fig. V.6), and rates of leaching declined to the levels in the control forest (Fig. V.5). Biomass accumulation in the burned plot continued to be higher than that in the control forest.

This study showed that, while there was some nutrient loss following the cutting and burning of the forest, the amounts lost were not high enough to decrease the productive potential of naturally occurring tree species.

6. Moderate intensity, small size, intermediate duration

(a)   Shifting cultivation

Shifting cultivation can be classified as a moderate intensity disturbance. The original vegetation is killed, but the soil is usually not so degraded that native plants cannot eventually re-establish.

Shifting cultivation is usually a small-scale disturbance. Ordinarily, shifting cultivators do not clear and plant more than a few hectares at a time. After crop production declines, the farmer abandons the plot and moves on to clear another one. The abandoned land and the vegetation that establishes on it is called fallow. 'Fallow' also refers to that period of time during which the land is not cultivated. Where population pressure is low and there is ample room and time for fallow vegetation to grow before another period of cultivation, shifting cultivation is indeed a small-scale disturbance. However, where the human population is increasing and there is a shortage of arable land, fallows become shorter and shifting cultivators can cumulatively cause a large-scale disturbance.

The active period of soil disturbance during shifting cultivation is usually only a few years. It is classified as a disturbance of intermediate duration, as opposed to selective logging operations which are short duration disturbances and as opposed to pastures which can be a disturbance of a decade or more.

There are dozens, if not hundreds, of variations of shifting cultivation practiced throughout the tropics. In the literature they are most commonly called 'slash and burn agriculture' or 'swidden cultivation', from the old English terminology. The practise is also referred to by a number of regional names, such as 'milpa' in Central America, 'chena' in Sri Lanka, and 'kaingin' in the Philippines (Nye and Greenland, 1960). In South America, the terms 'roza y tumba' and 'conuco' are frequently used (Watters, 1971). Many other terms are listed by Savage *et al.* (1982).

The many variations of shifting cultivation practises reflect diverse local conditions. Nutrient-rich soils can support nutrient-demanding crops, such as maize and sugar cane, while nutrient-poor soils can only support cassava (manioc) and such perennial tree crops as cashew. Seasonality and amount of rainfall influence the types of crop and the cultivation system used. Along the alluvial flood plain of the Amazon basin, cultivation is govered by the annual rise and fall of the river. Cultural factors also play a role. In Central America, the Ladinos (people of mixed blood) cultivate sugarcane and maize in the lowlands, and coffee and wheat at higher altitudes, while the Indians grow maize–bean–squash complexes at all altitudes (Watters, 1971). In parts of South-East Asia, 'mixed gardens' are common. These include fruit trees, bamboo, vegetables, and domestic animals. In Java, nutrients in organic waste are sometimes recycled into fishponds (Soermarwoto, 1977). The tools available for cultivation vary greatly. Near frontier villages in Latin America, it is possible to see natives using 'dibble sticks' to poke holes in the ground to plant seeds, while in a nearby plot soldiers on duty at the frontier are using tractors. The most common tool throughout the tropics is the long-bladed knife often called a machete or cutlass.

Despite the great variability in shifting cultivation systems, there are many commonalities among them with regard to patterns of nutrient cycling. In the next few pages, these common patterns of nutrient cycling and productivity during shifting cultivation in tropical forest areas are discussed.

(i) *Cutting and burning* The first step in site preparation is the felling of existing vegetation. Primary forest, secondary successional forest, or plantation forest are all sometimes used. In areas of primary forest, some of the biggest trees may be left standing, either because they are too difficult to cut down, or because they produce desirable fruit.

After cutting, trees and shrubs are allowed to dry for a period of several weeks to several months before burning. Important nutrient losses may occur during this interval. Sometimes logs are hauled off the site, and this can represent an important nutrient loss (Jordan and Russell, 1983). In a study of nutrient dynamics at an experimental site in Costa Rica, Ewel *et al.* (1981) found significant losses of nutrients due to tree harvest and leaching after cutting but before burning.

Burning can perform a number of functions: (1) it kills stump sprouts which would otherwise shade out crop species; (2) it kills some seeds of weedy vegetation that would also compete with crops; (3) it converts dried leaves and some wood to ash, thereby increasing the amount of nutrients immediately available for the first crop; (4) it removes most twigs and small branches which form a dense cover that makes planting difficult; (5) it reduces the cover available to insects, mammals, birds, and other crop consumers; and (6) it may also partially sterilize the soil, although microbial populations rapidly increase again (Nye and Greenland, 1960; Bennett *et al.*, 1974).

Burning is almost universally practiced, wherever climate permits. In continually wet areas, such as the lowland Atlantic coastal plain of Costa Rica, burning is sometimes impossible because there are insufficient consecutive rainless days to dry the cut vegetation.

During burning, nutrients in the vegetation are deposited in ash or are volatilized. The amount of the vegetation actually consumed by the fire depends on its dryness, the relative density (specific gravity) and diameter of the logs, whether the slash is piled or scattered, and on climatic factors. During an experimental burn of a secondary forest in Costa Rica, 30 percent of the initial amount of carbon, 22 percent of the nitrogen, and 49 percent of the sulfur were volatilized (Ewel *et al.*, 1981). Following the burn, some of the nutrient-rich

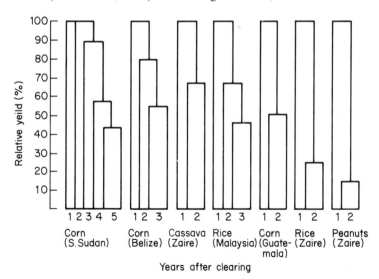

Fig. V.7. Decline in yield with continued cropping under shifting cultivation in forest environments. [Adapted from Norman (1979) with permission of University Presses of Florida.]

ash that had accumulated on the soil surface was either blown away or leached through the soil by the first rainfall.

In the wet tropics, where phosphorus is often the limiting factor, an important effect of the ash is to raise the soil pH. As a result, iron and aluminum become less soluble, and phosphates, complexed with aluminum at low pH, become soluble and available to crops (Chap. IV, section C).

(ii) *Nutrient dynamics during cultivation*   Because of the increase in available soil nutrients following burning, yields of the first crop almost always are higher than those subsequent crops (Fig. V.7), although there are occasional exceptions (Charley and McGarity, 1978). Because nutrients are more soluble following their conversion to ash, it has been suspected that nutrient leaching may be responsible for the decline in productivity during slash and burn agriculture. To test this idea, an experimental slash and burn plot was established in an Amazonian rain forest near San Carlos, Venezuela, and both primary productivity and nutrient leaching were measured in the experimental plot as well as in an undisturbed control forest (Jordan, 1985).

During the period of cultivation, total primary productivity declined from 5.6 t ha^{-1} a^{-1} dry weight in the first year to 4.1 t ha^{-1} a^{-1} in the third year, but edible yield dropped from 1.4 to 0.7 t ha^{-1} a^{-1} during the same interval. Net primary productivity of the control forest remained almost constant at about 11 t ha^{-1} a^{-1}.

Leaching losses of calcium, potassium, and magnesium in the slash and burn site were much higher than in the control site (Fig. V.8). While there was some phosphorus solubilization near the soil surface due to an increase in soil pH, phosphate leaching from the experimental site did not increase over that in the control plot (Jordan, 1985). Leaching of ammonium nitrogen also did not increase in the slash and burn site, but there was a significant increase in nitrate leaching after cutting and burning (Fig. V.8). The high amounts of nitrate leached from the cut and burned plot suggests that nitrification played an important role in increased nutrient loss. Nitrification is a process which results in nutrient loss following ecosystem disturbance in temperate regions (Vitousek *et al.*, 1979), and probably also causes nutrient loss during cultivation in the tropics (Nye and Greenland, 1960).

Nutrient dynamics in the whole ecosystem during the experiment are illustrated in Fig. V.9. The values for 1976 are those of the primary rain forest before cutting. The compartments are stacked, that is, added vertically. For example, for potassium, the bottom-most compartment is exchangeable potassium in the soil with a value of 23 kg/ha. Above this is potassium bound in humus, 53 kg/ha. The top of the humus compartment is plotted at 75 (23 + 52) kg/ha. Other compartments (see legend to Fig. V.9) are stacked above the humus, and the grand total of all ecosystem potassium is 378 kg/ha. For nitrogen, only total amounts, not exchangeable or available, are plotted for the soil.

The forest was cut in August 1976 and burned in December 1976. The loss of

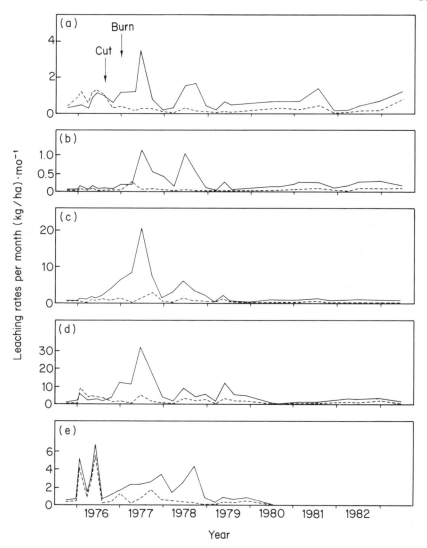

Fig. V.8. Leaching rates of nutrients from control (– – – –) and experimental (———) slash and burn agricultural plots in the Amazon Territory of Venezuela. (a) calcium; (b) magnesium; (c) potassium; (d) nitrogen as nitrate; (e) nitrogen as ammonium.

nitrogen from the ecosystem, shown just below the cut and burn arrow, was probably due to volatilization. Other nutrients showed no detectable losses during the cutting and burning.

During the burn there was a transfer of nutrients from the slash to the soil. This can best be seen in the calcium and magnesium graphs where the big

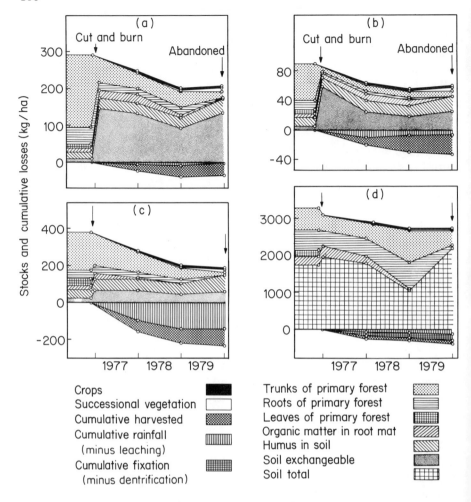

Fig. V.9. Stocks and cumulative losses of nutrients from a slash and burn agricultural site in the Amazon Territory of Venezuela: (a) calcium; (b) magnesium; (c) potassium; (d) nitrogen. The forest was cut in August 1976, and burned in December 1976. The plot was cultivated through the end of 1979, at which time it was abandoned.

increase in soil stocks is compensated by a decrease in the slash (mostly trunks of primary forest trees).

Starting with the burn, net losses from the slash and burn plot are plotted below the zero value on the vertical axis. Losses are due to leaching, harvesting, and denitrification.

There clearly are losses of calcium, potassium, magnesium, and nitrogen from the ecosystem as a result of cutting, burning, and cultivation. However, trends in the soil differ from those in the entire ecosystem. Total amounts in the soil increase after the burn. Calcium and magnesium remain high in the soil during the period of cultivation, despite leaching losses. This is because leaching losses from the soil are matched by nutrients moving down into the soil from ash and decomposing slash. Even potassium stocks in the soil do not decline during the period of cultivation. Total soil nitrogen stocks were variable during the period of cultivation, and were high at its conclusion in 1980. Burning increased the available soil phosphorus from 2 to 6 percent of total, but there was no detectable net loss of phosphorus from the soil (Jordan, 1985).

An important aspect of Fig. V.9 is that total soil nutrient stocks during cultivation did not decline markedly and remained above levels in the primary forest before cutting. These results contradict the previously accepted theory that slash and burn agriculture decreases soil nutrient stocks. As an example of how strongly the nutrient decline idea had been accepted, the following quote from Watters (1971) is presented:

> The most likely explanation of yield decline – declining soil fertility – was not apparent in soil chemical analyses carried out at intervals between 1933 and 1938 on Steggerda's [the reseracher's] four experimental plots. This result is so incredible, in view of the prolific literature on declining soil fertility on unfertilized cropped land in the humid tropics, that one is inclined to accept Cowgill's [the critic's] reassessment of Steggerda's data by means of calculating percentage changes in nutrient levels: she states that the results suggest variations in laboratory analytical techniques.

Nevertheless, recent investigators have found that, despite declining crop yields, nutrient stocks in soils of cultivated fields often remain higher than those in undisturbed forests. For example, in Ghana, on a 'reddish yellow latosol', Nye and Greenland (1964) found that exchangeable cations and Kjeldahl nitrogen increased after burning and then decreased, but that after two years levels were still above those which pre-dated forest clearing. Zinke *et al.* (1978), working in Thailand, studied the nutrient content of latosols in various stages of a 10-year rice–tobacco–cotton rotation system. They found that exchangeable calcium and magnesium were highest in recently cut and burned plots and that levels of these nutrients remained higher throughout the 10-year cycle than in a nearby 'old growth' forest that served as a control.

Studies in the neotropics show similar trends. Denevan (1971), working at slash and burn sites in eastern Peru, found that nutrient levels in the soil increased following clearing and then declined slightly. He questioned the importance of nutrient loss in the decline of crop productivity stating, 'In general, though nutrient levels declined somewhat, with probable corresponding decreases in crop yields, there is no evidence that this was the main reason for the abandonment of fields after only one or two years.' Harris (1971) examined a series of slash and burn sites in the upper Orinoco region of Venezuela. His comparison of the mean values of all forest soil samples with the mean values of all slash and burn soil samples showed that organic carbon, available phosphorus, and exchangeable calcium and magnesium all were higher in cultivated plots. Only exchangeable potassium and sodium were lower than in forest soils. Brinkmann and Nascimento (1973) studied changes in latosols in the central Amazon Basin during the first year of slash and burn agriculture. They found that exchangeable calcium and magnesium in farm plot soils were substantially higher than in soils of the undisturbed forest. Potassium was initially higher and then decreased to about the level of the forest. Total phosphorus was generally unchanged except in one flooded plot. Finally, on ultisols in the Amazon Basin of Peru, soil nutrients 10 months after forest cutting and burning were higher than in the pre-treatment forest (Sanchez, 1977).

One reason that declining productivity of crops could not be correlated with declining soil nutrient stocks in these studies is that availability of nutrients to crop plants was not distinguished from total soil nutrient stocks. Declining available nutrients, not total nutrients, may have been the critical factor. Although it is difficult to determine the exact proportion of nutrients that are 'available', soil pH can be a good index of availability of phosphorus (Fig IV.2). At the San Carlos site (Fig. V.9), crop productivity closely paralleled soil pH trends. Soil pH in the undisturbed forest was 3.8, went to 5.4 after the burn, and then gradually declined to 4.1 at the time of abandonment (Jordan, 1985). Soil pH also can regulate aluminum availability, and it is likely that this factor also played a role in crop decline.

Further evidence that it was the 'availability' of nutrients to crop plants and not a decline in total soil stocks which caused decline at San Carlos was that productivity of successional species in the abandoned slash and burn site was higher than that of the control forest (see next section). High productivity of successional species in the same plot where crop productivity was low suggests that it was not the total stocks of nutrients which were responsible for the decline, but rather the relative availability of the stocks to crop and weed vegetation. The successional vegetation was able to take up nutrients that were 'unavailable' to crop plants.

Disappearance of decomposing slash on the soil surface of the shifting cultivators' site also may influence crop productivity. Some of the products of organic matter decomposition may be highly effective in increasing the availability of phosphate fixed in the soil (Dalton *et al.*, 1952). Phosphorus availability is also related to the rate of mineralization of the phosphorus held in organic matter (Walbridge and Vitousek, 1984).

In montane sites, availability of nitrogen rather than phosphorus may be a factor in low crop productivity. Slow nitrogen mineralization (Chap. III, section D) may cause nitrogen limitation despite high levels of nitrogen in the soil.

In assessing the effects of slash and burn agriculture, it is important not only to distinguish between total and available nutrients in the soil; it is also important to distinguish between stocks of nutrients in the soil and stocks in the ecosystem. Although total soil nutrient stocks during cultivation often do not fall below the levels of pre-disturbance forest, total stocks in the entire ecosystem do decline. When ecosystem losses are distinguished from soil losses, it appears that the old theory which predicted nutrient loss during slash and burn cultivation *was* essentially correct. There is a loss of most nutrients from the ecosystem. The confusion arises because measurements of nutrient stocks in the soil do not reflect ecosystem losses. Stocks of nutrients in the soil may show relatively little change, because nutrients leaching out of the soil are replaced by nutrients leaching into the soil from ash and decomposing litter.

(iii) *Succession during cultivation and following abandonment* The weedy vegetation that establishes in shifting cultivation plots during their first few months of cultivation depends, in part, on the composition of the soil seedbank which survived the burn. Most of the germinating seeds are from primary and secondary tree species. Several times a year, the farmer will go through his field cutting out these seedlings and also the sprouts of trees not killed by the burn (Uhl *et al.*, 1982a).

After the first few weedings, the composition of the flora changes (Uhl, 1984). Sprouting stumps have exhausted their reserves, and die. The residual seed bank in the soil becomes depleted. The weed vegetation consists mainly of plants whose seeds have been transported into the site since the burn. Some of these are successional trees whose seeds are carried by wind, birds, bats, and other mammals. Herbs and grasses become increasingly important. Once herbs and grasses become established, they are more difficult to control than tree seedlings. Some of them complete their life cycle in a matter of months, and they increase exponentially. Rhizomatous grasses can spread vegetatively and, thus, more frequent weedings are required. Not only are more frequent weedings needed, but the chore is more difficult since cutting alone does not eliminate the grasses.

Although crop productivity declined as a function of time in the experiment at San Carlos described above, total net primary productivity increased after the abandonment of cultivation. The increase was due to the production of secondary successional species. Total primary production during the third year of cultivation was 4.1 t/ha. During the first three years after abandonment, annual production was 7.2, 11.4, and 12.4 t/ha. Meanwhile, the undisturbed control forest averaged 11.7 t ha[-1] a[-1] (including leaf litter) over the same time period (Jordan, 1985).

Although there are few data on total rates of primary production in abandoned slash and burn sites, studies reporting standing crops of successional biomass following abandonment (Table V.1) suggest that regrowth is not slow. Why is it that crop species decline in productivity while overall productivity increases?

An important reason may be that native species are able to utilize soil nutrients that are unavailable to crop plants. What mechanisms enable wild species to obtain nutrients that are unavailable to crop species, and to survive when crop plants cannot? Some of the mechanisms are the same as those discussed in Chapter II, while others may be specifically characteristic of successional species.

Table V.1

Above-ground standing crop biomass (in grams dry weight per square meter) at abandoned tropical slash and burn agricultural sites

Site location	Length of time since abandonment				
	2 years	4 years	5 years	6 years	7 years
San Carlos de Río Negro Venezuela	1280				
Izabal, Guatemala (Snedaker, 1970)	1419	2711	3667	4467	
Darien, Panama (Ewel, 1971)	1298	3796		4294	
Guarin, Columbia (Gamble and Snedaker, 1969)	1584	4839			
Belgian Congo (Bartholemew *et al.*, 1953)			7669		
Benin, South Nigeria (Nye and Greenland, 1960)				4609	
Veracruz, Mexico (Williams–Linera, 1983)					5268

Mycorrhizal fungi   Although mycorrhizae increase the ability of native species to obtain phosphorus and possibly other nutrients from the soil, many tropical crops also have mycorrhizae (Howeler *et al.*, 1982). The critical question, for which there is yet no answer, is whether mycorrhizae play a role in the competition for nutrients between native species and crops (St. John and Coleman, 1983).

Nutrient uptake kinetics   When soil nutrient availability is low, the rate of nutrient movement through the soil to the root surface can limit nutrient uptake (Nye and Tinker, 1977). Species with low nutrient uptake requirements can survive where nutrient diffusion rates are low, and will do better than species that require a higher rate of nutrient diffusion (Chapin, 1980). Most crop species have been bred for high nutrient uptake and high growth rates under conditions of high soil fertility. These species or varieties are at a disadvantage on low fertility soils when competing with species having low rates of nutrient absorption (Olson *et al.*, 1981; Chapin, 1983).

Root : shoot ratio   Wild plants growing in nutrient-deficient soil often produce a greater root biomass relative to shoot biomass than do crop plants and, thus, are better able to exploit infertile soils. For example, the root-

: shoot ratio of the principal crop species at the San Carlos site, *Manihot esculenta*, was 0.06, while the average ratio for successional species was 0.23 (Uhl, 1985).

Rhizosphere interactions   The greater capacity of wild plants to take up phosphorus in tropical soils of low nutrient availability may be due to the ability of their root exudates to stimulate decomposition of soil organic matter, hydrolyze organic phosphates, and dissolve rock phosphate. Nitrogen fixation also occurs among some wild species, although there are many cultivated crop and tree species which also form nitrogen-fixing associations.

Longer life span   The long life span of late successional and primary forest trees enables them to take up nutrients beyond their immediate need during seasons of nutrient abundance and store them for later use during seasons of nutrient scarcity (Chapin, 1980).

Leaf longevity   The leaves of late successional and primary forest trees are relatively long-lived (Mabberley, 1983). Long-lived leaves reduce the necessity for high nutrient uptake to replace leaves which are shed.

Efficiency of nutrient use   The lower nutrient concentrations in the tissues of late successional and primary species (Chapin, 1980) suggest that many wild species are able to produce a unit of biomass with fewer nutrients than can crop species.

Reproduction   Many species from infertile habitats do not produce a large seed crop every year. Often several years elapse between large seed crops. This reproductive pattern not only reduces nutrient use, it also keeps populations of seed predators at a relatively low level (Janzen, 1974).

Tolerance of acid soils   In acid soils with high concentrations of soluble aluminum and manganese, root development of crop plants can be inhibited. High aluminum concentrations in the roots impede the uptake and translocation of calcium and phosphorus, leading to nutrient deficiencies (Sanchez, 1976). Tropical trees that grow on acid soils have a high tolerance for aluminum and manganese (Baker, 1976).

Species diversity   Herbivory can cause a decline in crop production. There is little doubt that crop monocultures are more susceptible to pest outbreaks because of the short distances between host plants, and because herbivores are often host specific. Because successional communities are more diverse, attacks on native species often are less frequent or less damaging (Orians *et al.*, 1974).

Allelopathy   Higher plants synthesize substantial quantities of secondary compounds that can repel or inhibit other plants or herbivores. Differences in the quantity or quality of these substances may also contribute to the differential success of crop and successional species (Whittaker and Feeny, 1971).

As a result of these mechanisms, native successional vegetation is able to rapidly increase in biomass despite low soil fertility. The accumulation of biomass is accompanied by an accumulation of nutrients in plant tissues. This build-up of plant nutrients is the mechanism through which fallowing restores site fertility. The fallow vegetation takes nutrients which are unavailable to crops, and incorporates the nutrients in biomass. When the fallow vegetation is cut, the decomposing organic matter supplies nutrients in a form that crop plants can use.

(iv) *Patterns of succession*   The development of secondary successional vegetation varies greatly among sites, because of human factors (Maury-Lechon, 1982) and because of such site factors as nutrients and moisture (Ewel, 1980). The patterns also reflect the life cycles of plants (Gómez-Pompa and Vázquez-Yanes, 1974). When forest disturbance is chronic, species with short life cycles, such as grasses and herbs, are favored, but if the disturbance is short, tree species play a more important role (Uhl *et al.*, 1981, 1982a).

There appear to be distinct successional patterns in a particular area depending upon whether soil nutrients are depleted at the time succession starts (Chapin, 1980). When soil nutrients are relatively low, as, for example, after several years of cultivation without fertilization, the successional vegetation is characterized by high nutrient use efficiency, low nutrient uptake kinetics, slow growth, and the set of adaptations described under (iii), above.

In contrast, when abandoned fields are high in nutrients, the growth of early successional vegetation can be rapid, as it was in the studies reviewed by Bazzaz and Pickett (1980). The lack of mycorrhizae typical of early successional species on nutrient-rich sites (Janos, 1980a) may be a factor in the high growth rates achieved on such sites. When soils are relatively rich, the energy drain which mycorrhizae impose upon their hosts (Paul and Kucey, 1981; Odum and Biever, 1984) may be disadvantageous to the infected plants.

One neotropical genus typical of nutrient-rich successional sites is *Cecropia*. It grows rapidly in height and volume, and its tissue nutrient concentrations are high compared to those of other successional species or of primary forest species (Uhl, 1985). It is capable of rapid development without mycorrhizae (Mabberley, 1983). Genera with similar characteristics are *Musanga*, in Africa, and *Macaranga*, in South-East Asia. There apparently have been no studies relating establishment of these genera to soil fertility. However, observations on *Cecropia* throughout Latin America (C.F. Jordan, personal observation) suggest that the species of this genus establish most frequently on soils relatively high in nutrients, such as andosols in Central America, or on sites that have been cut, burned, and abandoned. They are less common and less vigorous on infertile soils such as oxisols and spodosols, and on sites that have been cultivated for a number of years before abandonment.

Another genus which apparently shares the characteristics of early successional species on nutrient-rich sites is *Phytolacca*. It is a genus of large herbs
which does not form any mycorrhizal association (Janos, 1980b). *Phytolacca*
often rapidly and densely colonizes tropical sites during the first year or two
following deforestation when soil nutrient levels are still high. For example,
Harcombe's (1977b) fertilized plots in Costa Rica and the irradiated site in
Puerto Rico, where leaf litter input was high, were heavily colonized by
Phytolacca. In contrast, nutrient-poor sites are rarely invaded by *Phytolacca*.

Following the establishment of early successional, nutrient-demanding species, soil nutrient content declines (Uhl and Jordan, 1984). Other species
become established which are able to maintain growth despite lower nutrient
concentrations. Their lower nutrient requirement, and perhaps their greater
shade tolerance, gives later successional and primary species a competitive
advantage over early successional plants. Consequently, the vegetative community changes (Fig. V.10).

The succession of species from highly nutrient demanding to less nutrient
demanding can be reversed by perturbations such as fire (Chapin and Van
Cleve, 1981). For example, a community may be dominated by species adapted
to low nutrient conditions. If a fire burns the litter of the forest, the ash forms a
nutrient-rich substrate to which fast growing, nutrient-demanding species such
as *Phytolacca* spp. and *Erechtites* spp. are well adapted, and these species
proliferate. As the pulse of available nutrients from the ash becomes depleted,
they die out.

Another very distinct type of succession occurs when grasses densely colo-
nize an abandoned site, inhibiting the establishment of trees. For example, in

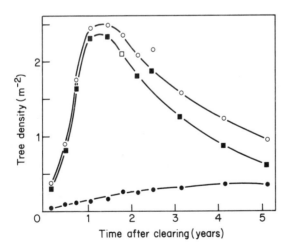

Fig. V.10. Density of secondary and primary trees in the first five years following
clearance in a forest reserve in Ghana. o—o, All species; ■—■, secondary species;
•—•, primary species. [Adapted from Swaine and Hall (1983) with permission of
Blackwell Scientific Publications, Limited.]

northern Thailand, during nine years of natural succession, dense stands of *Saccharum spontaneum* and *Imperata cylindrica* maintained an almost constant standing biomass of about 12 t/ha dry weight. Only after 20 years did woody vegetation become significant (Drew *et al.*, 1978).

(v) *Regrowth of leaves and fine roots* Rate of total biomass accumulation during fallow depends on soil fertility. However, recovery of leaf biomass appears less dependent on soils. At San Carlos, in the Amazon Basin, a successional site attained a leaf area index of 5.1 within five years, compared to 5.2 in the mature forest, (Uhl and Jordan, 1984). Golley *et al.* (1975), in Panama, and Bartholomew *et al.* (1953), in the Congo, also found that leaf cover approaching that in well-developed forest was reached within the first five to six years of succession. It took only 10 years, at a Nigerian site, for soil organic matter to reach levels approaching those in mature forest (Aweto and Areola, 1979).

Although total root biomass accumulation during succession may be slow (Berish, 1982), the growth of the fine, unsuberized roots responsible for nutrient absorption can be extremely fast. Just one year after a rain forest site in Costa Rica was cut, the total biomass of roots less than 2 mm in diameter was 92 percent of that in the undisturbed forest, and the biomass of roots less than 1 mm had reached 99 percent (Raich, 1980b).

(vi) *The Lua' forest fallow system* The discussion of slash and burn agriculture has shown that soil nutrient levels increase following cutting and burning of the forest. During cultivation, nutrient levels may decline, but usually not below the level of the primary forest. Following abandonment of cultivation, calcium, magnesium, and potassium levels in the soil continue to decline during succession, due to the accumulation of these nutrients in above-ground biomass (Jordan, 1985). It might be anticipated that these soil nutrients would continue to decline throughout succession, while ecosystem totals would increase due to the accumulation in above-ground biomass. It might also be anticipated that nitrogen, after an initial decrease during cultivation, would build up in the soil during succession due to increasing soil organic matter (Jordan, 1985).

The studies by Zinke *et al.* (1978) and Sabhasri (1978) of native cultivation systems on latosols in northern Thailand give a clear picture of nutrient stocks and dynamics during cultivation and fallow. Because these studies illustrate so well the interrelations of plant productivity and nutrient dynamics, the results will be given here. However, before doing so, it is worthwhile to quote Zinke *et al.*'s description of the cultivation/fallow cycle in order to clarify the cultural context of this agricultural system.

Burning the Fields
   The Lua' forest fallow system can be considered as a cycle lasting approximately ten years, beginning and ending in the early part of a dry season when the forest

Fig. V.11. Lua' forest fallow cultivation cycle of 10 years. Dates show other years when the currently cultivated plot (1967) was cut and burned. Solid arrows indicate soil and vegetation sample times relative to phases of the cycle. Dashed arrows indicate additional sample times. [Adapted from Zinke *et al.* (1978) with permission of Honolulu University Press of Hawaii.]

cover is cut in a previously cultivated field [Fig. V.11]. The trees and other vegetation are felled to lie uniformly; they form a fuel bed which is allowed to dry for six or more weeks. Farmers fell smaller trees, leaving a one-half-meter or one-meter stump from which sprouts may later grow. These stumps show evidence of repeated cutting and coppice growth, confirming the fact that the fields have been used repeatedly. Lua' farmers leave most of the larger trees standing, but trim their branches to reduce shading of the crops.

The fields are burned a few weeks before the end of the dry season. Too late a burn may result in a fuel bed wet by thunderstorms that occur several weeks before the start of the monsoon, while too early a date may find fuel incompletely dried. The approximate date of burning is known several months in advance, and in 1968 the fields were burned on the preselected date of March 25, in coordination with the burning of fields of an adjacent village. Lua' avoid burning during a time of waning moon for fear there will be too many weeds. A leeway of several days is possible if the headman thinks the weather is unfavorable or if a neighboring village plans to burn their fields at an earlier date.

Fuel breaks are prepared around the area to be burned to prevent the escape of fire into the adjacent uncut forest. At sunrise on the day of burning, the oldest women go to the edge of the village and sacrifice a little rice and other foodstuffs, liquor, yarn, and cowrie shells to the ancestral spirits. Later in the morning older men who are religious leaders construct an altar of branches and leaves in the forest along the trail to the year's fields. They sacrifice a dog and pig and several chicks to the field spirits, calling for their assistance in achieving a good burn. The old men continue the ceremony until youths and young men light the fuel bed, offering small amounts of rice liquor to the spirits and drinking the remainder from the sacrificial bottles. The fire-setters maintain strict sobriety for their dangerous task.

The fields have a considerable range of elevation with some steep slopes (1967 range from 680 to 820 meters above sea level; 1968 range from 740 to 1,040 meters, with slopes up to 70 percent) creating the potential for strong updrafts. The torch-bearers are fully aware of the danger of being caught in the rapid spread of flames sweeping up the slopes. At about midday, when conditions are hottest and driest (32°C and 10–20 percent relative humidity at the time of the 1968 fire), they begin to light the fuel bed, starting on the ridges.

After lighting the highest points on the fields, the men run down along the field edges, and the fire begins to burn down the slope. As the fuel begins to burn, it causes an indraft and by the time the fire reaches the lower edges of the fields, a strong wind is blowing into the fire. In 1968 there was between 5 and 6 km of periphery to be lit. The firing of the lower periphery was accompanied by indraft winds of 40 to 65 km per hour. Small, intense firestorm whirlwinds developed in the fuel bed area as it burned. These were spiral masses of flame up to 100 meters tall, developing intense winds at their centers. They were strong enough to tear out temperature measuring templates partially buried under some of the slash.

In the 1968 burn, approximately 54,636 metric tons (oven dried weight) of fuel was burned on 94.2 hectares in less than an hour. One effect was a 10–20,000-foot column of smoke and ash, topped by a cumulus cloud with precipitation downwind. Flocks of swallows flew through the edges of the smoke cloud, hunting insects in the updrafts.

The burn left a fairly uniform ash bed laid over the fields. Most leaves, branches, and smaller stems were burned. In the next few weeks, the villagers laid charred logs horizontally on the slopes behind stumps to form revetments which are effective in controlling soil creep. They placed as many as 100 logs per hectare, and collected the rest of the larger logs and stems for field border markers, fencing, and firewood. The remainder of unburned slash they piled and reburned in small hot fires. The ash is black over most of the area, with whiter ash where fire temperatures were higher and small regions of white ash and reddened earth where the reburn piles had been.

Villagers plant rice in the ash-covered fields after cleaning up the unburned slash. Men jab the soil with long metal-tipped bamboo poles making about 14 holes per square meter. Women, children and older men follow, casting a few seeds at each hole.

The rice begins to grow with the onset of the monsoon rains, and is harvested toward the end of the rainy season. Even before the rains begin, soil is moist below about 5 cm, because the early cutting of the forest conserves soil moisture which would otherwise be lost by evapotranspiration in the dry season.

Despite continuous weeding throughout the rice-growing season, herbaceous species, dominated by *Eupatorium odoratum*, cover the swiddens soon after harvest. Most of the trees are not killed by cutting or the fire. They begin sprouting from stumps soon after the fire, and eventually form a coppice of sprout growth shading out most of the herbaceous plants. Thus the fallow fields are eventually dominated by tree species, with some bamboo. A high proportion of presumably nitrogen-fixing leguminous trees occur in the forest fallow vegetation, which grows for about nine years before the cycle is repeated [Fig. V.11]. [Reprinted from Zinke *et al.*, 1978, by permission of the East–West Center, University of Hawaii.]

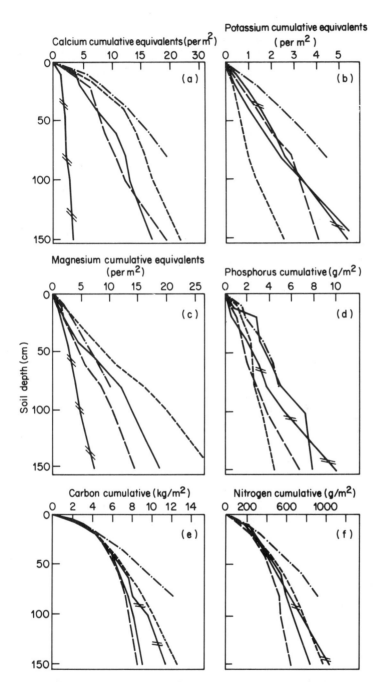

Fig. V.12. Cumulative amounts of soil nutrients and carbon with increasing depth, at various stages of Lua' cultivation/fallow cycle: – – – –, current burned field; —.—, one year after burning; –··–··–, four years after burning; ——, seven years after burning; ——, old unburned forest. [Adapted from Zinke *et al.* (1978) with permission of Honolulu University Press of Hawaii.]

Nutrient dynamics during the cycle are reflected by stocks of soil nutrients at various stages in the cycle (Fig. V.12). These graphs represent cumulative amounts of nutrients with increasing depth in the soil. The recently burned field had slightly lower amounts of calcium than did the field one year after burning (Fig V.12(a)), probably because calcium in the ash had not yet leached down into the soil. Four years after burning, the soil calcium level had decreased because calcium had been incorporated into the above-ground vegetation. Between four and seven years there was little change in the soil, perhaps because there was little biomass accumulation, at least above ground (Zinke *et al.*, 1978). If the area had been abandoned indefinitely, calcium uptake by vegetation would again have increased, and soil calcium levels would have decreased, approaching the level in unburned forest.

Magnesium (Fig. V.12(c)) showed a pattern similar to that of calcium except that the recently burned plot had greater amounts than did the one year old field. Like calcium, the bulk of the magnesium in the mature forest is in the wood and, consequently, soil levels are low.

Potassium (Fig. V.12(b)) was relatively low in the soil the year of the burn (current field), probably because the ash had not yet washed down. First year levels were high. Potassium then decreased until the fourth year, held steady until the seventh, and then decreased during the last three years.

Available (reagent extractable) phosphorus (Fig. V.12(d)) increased between the burn and the first year as ash leached down into the soil, raising the pH and increasing the solubility of phosphorus complexed with iron and aluminum. Like calcium, soil phosphorus then decreased until the fourth year. Between four and seven years, phosphorus increased again, perhaps because of an interaction with increased root growth during that period. Available phosphorus in the fallow did not differ greatly from that in the old unburned forest, because total (acid digest) soil phosphorus stocks probably never changed very much. This is so because only a small proportion of the total phosphorus in rain forest ecosystems is contained in the vegetation (Fig. II.4(b)). Most is fixed in the soil and is affected only slightly by biomass dynamics.

Carbon and nitrogen showed similar trends (Fig. V.12(e) and (f)). As slash fragments were incorporated into the soil during the first year, soil carbon and nitrogen content increased. Then, as organic matter decomposed over the next three years, soil carbon and nitrogen levels declined. After the seventh year, carbon and nitrogen increased again. At the end of the fallow, they were close to levels in the old, undisturbed forest.

In summary, this study illustrates the dynamics of soil nutrients during cropping and fallowing. The crop plants take up the nutrients made readily available by cutting and burning. As nutrient levels decline, fallow vegetation comes in. The successional vegetation continues to take up nutrients, and causes a continued decline in soil calcium, magnesium, and possibly potassium stocks. However, the successional vegetation causes an increase in soil organic matter and, thus, soil carbon, nitrogen, and phosphorus increase during the fallow.

The Lua' system of cultivation may be sustainable in the long term without fertilization, since nutrients apparently reaccumulate sufficiently for cropping every 10 years. However, in many tropical areas the cycle of cultivation and fallow is not closely regulated or followed. Often, the cycle is interrupted by social, political, or economic forces beyond the control of the native cultivator or local tribe. When this happens, long-term sustained yield may not be possible.

(b) Other disturbances due to management

Slash and burn agriculture has been given as an example of forest disturbance of moderate intensity, small size, and intermediate duration. There occur now, or have occurred in the tropics, many other management systems for the production of food and fiber crops, and that also fall under this category of disturbance. They are practiced in order to maintain ecosystem productivity with a minimum of soil disturbance, or to restore soil fertility before irreversible degradation occurs.

Weaver (1979) recognized eight major categories of management:

(i) *Shifting cultivation*   In shifting cultivation, an area is cut and burned (if possible) and crops desirable for subsistence or exchange are cultivated for a few years. When productivity declines, the plots are abandoned and soil fertility is restored by the vegetation which naturally invades the site. Often, fruit trees or other trees valued for their wood are deliberately planted before abandonment and these contribute to increasing soil fertility as well as yielding useful products. This system is practical only where population density is low.

(ii) *The corridor system*   This system was developed in densely populated regions of Africa to give more order and control over land use than is possible with shifting cultivation. Farmland was divided into strips which were cultivated and then fallowed sequentially. A 12-year fallow was followed by two seasonal crops, an annual crop, and a perennial crop. As population pressures increased, the system was abandoned.

(iii) *Taungya*   This system was initiated in Burma in the 1860s on public lands. It is a cooperative system between local farmers who plant food crops simultaneously with the timber species desired by the government's forestry operation. The farmer can get one or two crops before the tree canopy closes and annual cropping must be abandoned. Annual or perennial crops growing between tree seedlings hold nutrients and prevent erosion until the trees are large enough to stabilize the soil (Bruijnzeel, 1982).

(iv) *Tree intercropping*   In this system, the economically desirable crop is often an understory tree species, such as coffee or cacao, planted beneath larger trees, sometimes nitrogen fixers, which sustain soil fertility and provide shade. The same plot of land is cultivated indefinitely. Nutrient restoration may be carried out by the overstory trees.

(v) *Simulation of natural succession* In this system, species which establish naturally following intensive cultivation are replaced by analogous species with greater economic value. A hypothetical sequence might consist of: (a) annuals and plants whose roots and stems are harvested, (b) bananas and plantains, (c) palms, and (d) cacao, rubber, or commercial timber species.

(vi) *Self-sufficient farms* These farms completely recycle nutrients within the management unit. Many of the units contain livestock, fruit and forest trees, home gardens, pasture, ponds stocked with fish, and fields of multiple crops. Animal wastes are washed into a pond whose waters are used to irrigate crops. Plant residues and forage are employed as animal feed or green manure. A variation of this is the 'chinampas' system of Mexico (Gliessman *et al.*, 1981) in which runoff from elevated plots fertilizes aquatic weeds in surrounding canals. The weeds are then harvested as organic fertilizer for the plots (Quiroz Flores, 1980).

(vii) *Scattered or row trees* Trees planted in rows between cultivated fields or along property boundaries help prevent wind erosion in dry areas and erosion by running water in wet areas. Trees can be used as living fenceposts; that is, trees are planted along pasture boundaries and wire fencing is fastened to them. In some regions, trees that fix nitrogen are planted in and around pastures to maintain the nitrogen supply for the grasses.

(viii) *Forest blocks* Upper slopes and hilltops that are left undisturbed are beneficial for downslope cropped fields. They help prevent erosion, and leaf litter from the upslope forest improves the organic matter content of the cropland. In addition, when the cropland is abandoned, seeds from the upslope forest readily invade and forests can re-establish quickly. Some rice paddies, productive for centuries in the Philippines, are interspersed among parcels of forest. The forests have been maintained for religious reasons (P.B. Sears, cited in Weaver, 1979).

Eight major categories of tropical land management have been listed, but there are virtually hundreds of systems which differ in the crops planted, rotation time, cultivation methods, and many other details. Management systems in which trees are grown simultaneously with herbaceous crops are termed agroforestry or agri-silviculture. Even within the general category of agroforestry, there are a wide variety of methods.

An idea of the breadth of tropical agroforestry can be gained from the publications of international agroforestry research centers, such as the ones in Turrialba, Costa Rica (CATIE, or Centro Agronómico Tropical de Investigación y Enseñanza) and Nairobi, Kenya (ICRAF, or International Council for Research in Agroforestry). Several important publications are: *International Cooperation in Agroforestry* (Chandler and Spurgeon, 1979), *Soils Research in Agroforestry* (Mongi and Huxley, 1979), *Agro-forestry Systems in Latin America* (De las Salas, 1979), and *Agro-forestry in the African Humid Tropics* (MacDonald, 1981).

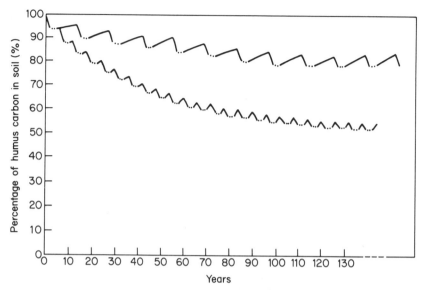

Fig. V.13. The approach of humus carbon in the soil to equilibrium conditions under shifting cultivation with a two-year crop and 12-year fallow cycle (upper curve), and a two-year crop and four-year fallow cycle (lower curve). No allowance is made for erosion losses. [Adapted from Nye and Greenland (1960) with permission of the Commonwealth Agricultural Bureaux.]

7. Disturbances of moderate intensity, variable size, and long duration.

(a) Savannazation

In disturbances of moderate intensity there is little long-term decrease in the soil fertility, when increasing stocks of soil nutrients during the fallow period compensate for losses during cultivation and harvesting of crops. When the fallow period is too short to completely restore the soil fertility before the next cropping cycle, and the productive capacity of the soil gradually declines, the disturbance can be considered as long term. Nye and Greenland (1960) illustrated this with theoretical calculations of the decrease in soil humus with increasingly short fallow periods (Fig. V.13). The calculations are based on biomass accumulation rates and decomposition of soil organic matter during cultivation.

Because the soil nitrogen stock is diminished only slightly during one cultivation cycle (Fig. V.9), nitrogen does not seem to be a critical factor when primary forest is converted, and cultivation is interrupted by long periods of fallow. However, when the fallow period is relatively short, as are those shown in Fig. V.13, the soil nitrogen stock gradually decreases and nitrogen can become limiting. Since soil organic matter is the principal form of nitrogen storage, patterns of nitrogen decrease follow carbon patterns in Fig. V.13.

Other factors, besides fallow length, play critical roles in the nutrient dynamics of disturbed tropical sites. One factor is the length of the dry season and another is fire.

Where the dry season is relatively long, the wet and moist forests give way to savanna (Table V.2), and annual net primary production is lower. Lower annual rainfall and lower stocks of biomass, carbon, and nitrogen result in lower annual rates of soil respiration and, hence, lower rates of carbonic acid production (see Chap. I). Consequently, acidification and leaching are reduced. Stocks of calcium, potassium, and magnesium in the upper soil horizons are higher than under rain forest, and, because of higher soil pH, phosphorus is more available and less limiting to plant growth.

However, the lower rate of biomass accumulation in these drier regions leads to a low rate of nitrogen replenishment during fallows (Fig. V.14). Repeated cycles of cultivation and short fallows diminish nitrogen stocks relative to stocks of calcium, potassium, and magnesium. Furthermore, the nitrogen present is often unavailable to plants. Due to lower bacterial activity under dry conditions, the nitrogen in decomposing litter and soil organic matter is mineralized slowly in savanna areas. Evidence that nitrogen is often limiting in savanna soils comes from fertilization experiments. Crops, such as cereals, yams, and cotton, showed greater growth response to nitrogen than to phosphorus or potassium (Nye and Greenland, 1960).

Fire is another factor critical in vegetation dynamics during disturbance. Fire plays a very important role in the formation of tropical savannas. According to Nye and Greenland (1960):

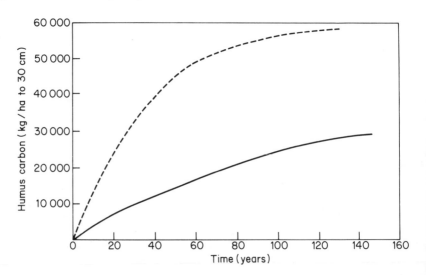

Fig. V.14. Hypothetical increase of humus carbon in tropical forest (– – – –) and savanna (———) starting with a bare mineral soil and production and decomposition rates typical of the respective regions. [Adapted from Nye and Greenland (1960) with permission of the Commonwealth Agricultural Bureaux.]

Table V.2

Arrangement of climax and secondary vegetation formations, with special reference to Africa: overlap in rainfall formations is caused in part by annual distribution patterns*

Rainfall (mm/year)	Climatic climax formations	Secondary formations		
		Mainly woody	Mainly grass	
Over 1650	Moist evergreen forest	Secondary bush, secondary forest	*Imperata cylindrica* (Asian var.)	
1140–2300	Moist semi-deciduous forest	Secondary bush, secondary forest, semi-deciduous thicket	High grass savanna – *Imperata cylindrica* var. *africana* *Pennisetum purpureum*; *Andropogon* spp.; *Hypharrhenia* spp.	
900–1650	Dry deciduous forest	Deciduous thicket, woodland	Tall bunch-grass savanna	
Under 1000	Deciduous thicket		Short bunch-grass savanna	

* Reproduced by permission of the Commonwealth Agricultural Bureau from P.H. Nye and D.J. Greenland (1960). The soil under shifting cultivation. Technical Communication No. 51.

In Africa much of the land now under high-grass savanna was once under moist semi-deciduous or, rarely, moist evergreen forest. The high-grass savanna is dominated by very tall coarse grasses from 2 to 5 meters high at flowering. The species are members of two tribes, the Paniceae containing notably *Pennisetum purpureum* (elephant grass) and *Panicum maximum* (Guinea grass), and the Andropogoneae, containing notably *Imperata cylindrica* (not always a 'high' grass, often being under 2 meters), *Andropogon gayanus*, and *Hyparrhenia* spp. The high-grass savanna forms the fire climax in regions where the rainfall exceeds about 1100 mm per year and the dry season lasts up to 6 months.

The Paniceae tend to be dominant in moister areas with short dry season; for example, elephant grass, the tallest and most productive of African grasses, is extensively developed in southern Uganda. It also occurs in parts of West Africa immediately to the north of the present forest boundary. The Andropogoneae dominate a much larger area, adjacent to the high forest. In the northern hemisphere they stretch across the continent from the Atlantic Coast of Guinea, in the west, to the upper reaches of the White Nile in the east; and in the southern hemisphere they dominate the southern half of the Belgian Congo.

In tropical America man, aided by fire, has caused far less destruction of the original vegetation than in Africa, and the degraded forest type of habitat is less common. True high-grass savanna has not yet been recognized, though *Imperata braziliensis*, the American analogue of *Imperata cylindrica*, does occur in over-cropped land. In Asia, large areas of the Philippines and Indonesia, and smaller areas in other parts of the former moist-forest region, have been replaced by 'cogonales'. The dominant grass is *Imperata cylindrica* var. *major*, and *Saccharum spontaneum* is frequently mentioned. Unlike the African variety of *Imperata cylindrica*, which is displaced in a few years by other grasses easier to clear for cultivation, the Asian form is the fire climax, and these cogonales are useless to the cultivator. If protected from fire they will eventually return to forest.

The grasses of the tall-bunch-grass savannas are about 1 meter high at flowering, and are dominated in the climax stage by the Andropogoneae. Trees and shrubs vary greatly in density, according to the intensity of cultivation, burning, and grazing, and the type of soil, ranging from almost total absence to a closed canopy of woodland. In Africa, the tall-bunch-grass savanna of the nothern hemisphere lies between the high-grass savanna and the short bunch grasses and semi-desert grasses and scrub on the fringe of the Sahara. In the southern hemisphere it covers much of the Rhodesias, and dominates East Africa from Zululand to Zanzibar. It grades into the short-bunch-grass savanna at a rainfall of about 900 mm.

In America, tall-bunch-grass savanna, entirely analogous to the African form, is widespread, notably on the Brazilian plateau, the Guiana highlands, and the Orinoco plains.

The short-bunch-grass savannas are about 0.3 meters high at flowering. The short grasses and scattered flat-topped *Acacia* trees are a familiar sight in the northern and southern hemispheres of Africa under rainfalls of 500–900 mm. Much of this region is too dry for successful rain-land cultivation. In tropical America and Asia short-bunch-grass savanna is very restricted since areas with rainfall under 900 mm per year are limited. [Reprinted from Nye and Greenland, 1960, by permission of the Commonwealth Agricultural Bureaux.]

The dynamics of grassland vegetation and of the microorganisms which control nitrification and denitrification following fire are shown in Fig. V.15. Increased bacterial activity following a fire results in nitrogen mineralization which in turn causes increases in grass productivity.

The dynamics of soil processes following fire are shown in Fig. V.16. The

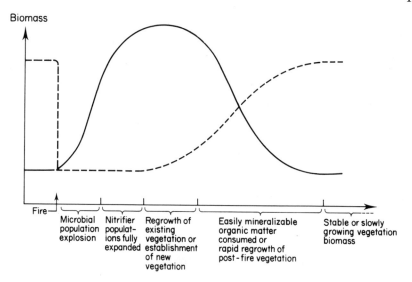

Fig. V.15. Relative time sequence of hypothetical responses of organisms to fire: ——, microorganisms; — —, vegetation. [Adapted from Woodmansee and Wallach (1981).]

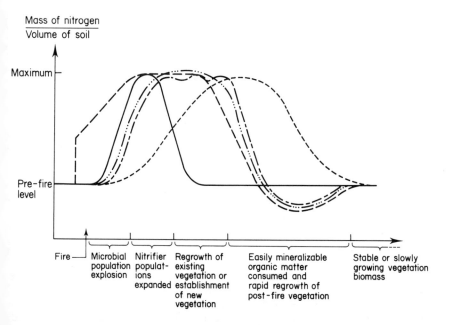

Fig. V.16. Relative time sequence of nitrogen transformations in soils ——, NO_2 and chemical dentrification; — — —, NH_4 and NH_3 volatilization; – – – –; nitrogen fixation; —...—, NO_3; — – —, biological dentrification. [Adapted from Woodmansee and Wallach (1981).]

126

first change is volatilization of ammonium. After the fire, which either kills or damaged most plants, competition for nitrogen is temporarily decreased and the activity of nitrifying and denitrifying bacteria increases. The increase in soil pH due to ash deposition may also increase microbial activity, resulting in higher losses due to nitrate leaching and denitrification. As native vegetation re-establishes, nitrogen fixation begins. At the same time, an increasing proportion of the ammonium in the soil is taken up by rapidly growing higher plants. Consequently, the substrate for nitrifying and denitrifying bacteria is reduced and their activity declines. Stocks of nitrogen in the soil again increase.

When fallow cycles are short or when the incidence of fire is frequent, net primary production is relatively low and, consequently, the uptake of nitrogen from the soil is low. This should provide more substrate for nitrifying and denitrifying bacteria and stocks of soil nitrogen should gradually decrease. Field studies have shown that stocks of soil carbon and nitrogen are, in fact, proportional to fallow time (Greenland and Nye, 1959).

Figure V.17 shows how increased fire frequency can convert woodland into

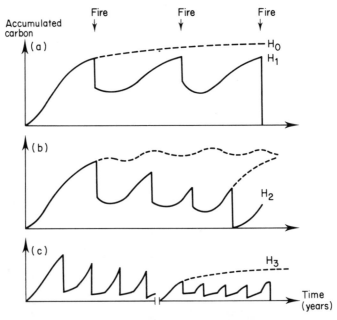

Fig. V.17. Theoretical patterns of carbon accumulation in total live plant biomass under various frequencies of burning. In graph (a), the dashed line (H0) represents dynamics under conditions of no burning, and the solid line (H1), burning separated by long intervals. In graph (b), the solid Line (H2) represents dynamics under intermediate intervals of burning. The dashed line represents stocks to which the ecosystem could recover after any particular burn, if fire were supressed after that burn. Intermediate burning does not appear to inhibit full recovery of carbon stocks. In graph (c), the solid line represents dynamics under frequent intervals of burning, and the dashed line (H3) represents the hypothesis that frequent fires eventually lower the capability of stocks of carbon to recover to pre-burn levels. [Adapted from Olson, 1981]

savanna or grasslai. ~~~~ en fires are infrequent, trees are an important successional compone. ~~~ llowing disturbance; biomass builds up to levels almost equal to those under completely undisturbed conditions (Fig. V.17(a)). When clearing is more frequent (Fig. V.17(b)), there is insufficient time for a full recovery of soil carbon and nitrogen stocks. However, if disturbance ceases, the potential for full recovery still remains. If burning is very frequent, (Fig. V.17(c)), woody vegetation cannot become established. Once the cycle of frequent fires begins, the process of savannization accelerates. Fire favors the establishment of grasses because it does not kill their underground rhizomes which survive and sprout at the start of the next rainy season. In contrast, fires tend to kill woody seedlings which generally lack substantial underground carbohydrate reserves. As grass cover increases, conditions for a more complete and thorough burn improve, because fire can spread readily through the accumulating grass litter. Moreover, dense grass cover also inhibits the establishment of tree seedlings.

Total ecosystem nutrient stocks are reduced in forest areas converted to grassland by repeated burning. Whether nutrient depletion at such sites prevents the re-establishment of forest vegetation is a subject of controversy. Figure V.17(c) suggests that irreversible changes can occur. However, other data showing that some savanna soils are more fertile than nearby forest soils have been taken as evidence that lack of nutrients will not inhibit recovery of savanna to forest (Budowski, 1956).

A study of biomass and nutrient stocks in burned and unburned successional areas in the Gran Pajonal of Peru (Scott, 1978) helps to clarify the dynamics of ecosystems under long-term fire stress. The Gran Pajonal, in the Andean foothills of western Amazonia, was originally forested, but is now mostly covered by grasses and scattered shrubs which have invaded former shifting cultivation sites.

At the time of abandonment, the fields are dominated by the fern *Pteridium aquilinum*. If no burning occurs, the *Pteridium* is soon overtopped by a young secondary forest dominated by *Cecropia* spp. and other early successional trees up to 15 meters tall. This stage lasts about 10 years. These trees are replaced by a mid-successional forest up to 20 meters tall, that changes, at about 50 years, to a mature secondary forest, and finally to primary forest with a canopy at 35 meters.

If burning is frequent following the abandonment of farm plots, the succession of communities is quite different. Fire suppresses the tree species, and favors *Pteridium*, certain grasses, and sedges. This is locally called the 'chac-chac' stage. The chac-chac gives way to a cover dominated by *Imperata brasiliensis*. Soil erosion increases, infiltration decreases, and there is invasion by *Andropogon* spp. Frequent burning results in maintenance of the *Andropogon*-dominated grassland.

The change in above- and below-ground nutrient stocks during fire-free succession in the Gran Pajonal is shown in Fig. V.18. The trend towards increasing ecosystem nutrient stocks but decreasing soil stocks is shown very

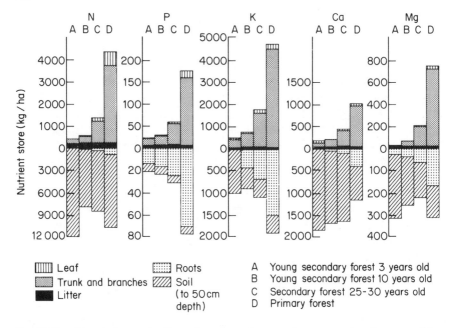

Fig. V.18.   Vegetation and soil nutrient stores in secondary successional communities. [Adapted from Scott (1978) with permission of the Department of Geography, University of Hawaii at Manoa.]

clearly. Although increasing amounts of below-ground phosphorus, potassium, calcium, and magnesium are shown as succession progresses, most of these below-ground stocks are in root tissue. The amounts actually stored in the soil clearly decline.

Nutrients in the fire disturbed Gran Pajonal ecosystem are shown in Fig. V.19. The stocks of some soil nutrients are lower in the fire-disturbed sequence than in the non-disturbed sequence. On the burned sites soil nitrogen ranges from 5000 to 8000 kg/ha, while in the undisturbed sites (Fig. V.18) values range from 8000 to 11 000 kg/ha. Above-ground nitrogen decreases with burning, whereas it gradually increases on unburned sites.

There is little difference in available soil phosphorus between burned and unburned sites. Since available (soluble) phosphorus is rapidly fixed in highly weathered soils of the Amazonia, total phosphorus stocks in the unburned and burned sites should be similar. The biggest difference between the burned and unburned sequences is in potassium. In the burned area, there is less than 200 kg/ha of potassium in the soil, while in the unburned area it ranges from 400 to 1000 kg/ha. However, soil stocks of calcium and magnesium are higher in the burned site than in the unburned successional and primary forests.

Total ecosystem nutrient stocks are relevant to the question of forest regrowth on savanna sites. The 'old grassland' (Fig. V. 19(c)) has a total about 8000 kg of nitrogen per hectare, while the primary forest (Fig. V.18(d)) has

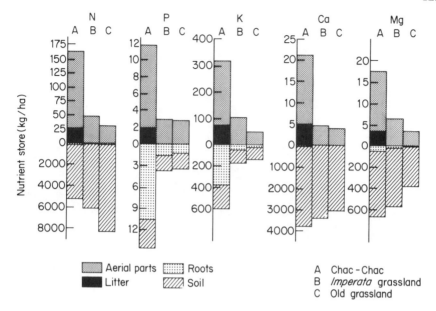

Fig. V.19. Vegetation and soil nutrient stores in a fire sequence, showing fern community (chac-chac), *Imperata* grassland, and old grassland. [Adapted from Scott (1978) with permission of the Department of Geography, University of Hawaii at Manoa.]

about 15 000 kg/ha when above- and below-ground stocks are added. Old grassland has about 5 kg of phosphorus per hectare, while primary forest has about 250 kg/ha. For potassium, the differential is 150 versus 6500 kg/ha, and for magnesium, 400 versus 1000 kg/ha. Only calcium stocks are approximately the same at both sites, perhaps because the underlying rock is calcareous shale and limestone.

In order for forest vegetation to re-establish on these degraded savanna sites, the total stocks of nutrients must be rebuilt. The stock of potassium must be rebuilt from 150 kg/ha to about 6500 kg/ha. The input of potassium in rainfall was estimated to be 5.23 kg ha^{-1} a^{-1}. If 100 percent of this input were taken up by the vegetation and there was no input from soil weathering, it would take $\dfrac{(6500 - 150)}{5.23} = 1214$ years for the full stock of potassium to be regained. If the primary forest at this site ever recovers, it will be a slow process.

(b) Large-scale spraying of herbicides

Between 1961 and 1971, during the Vietnam War, over 72×10^6 liters of herbicides were sprayed from the air onto the forests of southern Vietnam (Westing, 1984). The long-term effects of herbicides on the forest were the subject of controversy during a conference in Vietnam during January 1983

130

Fig. V.20.   Photograph of upland rice field established under defoliated forest, Vietnam, 1983.

(Carlson, 1983). Some delegates claimed that the ecological damage produced by spraying with herbicides may become spontaneously worse with time. For example, areas denuded of vegetation may suffer erosion or other deleterious transformations, or they may be invaded by noxious plants, such as *Imperata*, which impede regrowth of the original flora (Van Trung, 1984). Such subsequent damage, however, may have been caused by peasant farmers moving into sprayed areas to cultivate fields (Zinke, 1984). Peasants quickly moved into herbicide-sprayed areas because leaf litter fall supplied a pulse of nutrients to the soil (Van Huay, 1984) and the defoliation allowed sunlight to penetrate the canopy. It was much easier to initiate cultivation in such areas than in areas where the trees had to be cut by hand. Crops were planted directly beneath the decomposing trunks (Fig. V.20). When crop productivity declined, they burned the fields to encourage the growth of grasses for cattle grazing (Ashton, 1984). Repeated burning favors the growth of commercially worthless grasses, including bamboo, and a sequence of ecological changes resembling savannization occurs. Frequent burning may be the factor which prevents the re-establishment of forest vegetation, (Norman, 1983; Galston and Richards, 1984).

In areas that had been sprayed but not further disturbed, recovery was similar to that which occurs following heavy logging in the region. Rapidly growing secondary successional species invaded and formed a canopy by January 1983 (Ashton, 1984). Beneath this canopy, seedlings of primary forest trees were abundant, and if there is no further disturbance, primary forest species should again be dominant in less than a century (E. Brunig, personal communication).

The importance of fire and other disturbance in preventing recovery following herbicide treatment is indicated by an experiment in Puerto Rico, where further disturbance of herbicide-treated plots was prevented. Natural forest regeneration occurred rapidly and succession was similar to that following other types of disturbance (Tschirley, 1969; Dowler and Tschirley, 1970).

In coastal mangrove swamps whose soils contain a high concentration of pyrite (FeS_2), herbicide spraying had a damaging effect. The herbicides killed mangroves (*Avicennia* sp. and *Rhizophora* sp.) (Snedaker, 1984) and, consequently, large areas of mud flat were exposed to the drying influence of direct sunlight. Under these conditions, the pyrite is oxidized, forming sulfuric acid (Stumm and Morgan, 1981). This makes vegetation re-establishment considerably more difficult. However, at least in certain areas visited by conference participants in 1983, workers had replanted large areas that had been barren in 1972, and the vigor of the plants (Fig. V.21) left the impression that reestablishment would be successful.

(c) Continuous cropping with fertilization

Since the early 1970s, a series of experiments has been carried out near the foothills of the Andes in eastern Peru to determine the feasibility of continuous cropping with fertilizer application as an alternative to shifting cultivation (Villachica et al., 1976; Villachica and Sánchez, 1978). With increasing population, less land is available for shifting cultivation, periods of fallow must be shortened and, as a result, crop productivity declines.

The results of these experiments have shown that fertilization improved yield and prolonged the period of good yield (Sánchez et al., 1983). Nevertheless, even fertilization failed to halt yield declines over a period of several years. For example, Fig. V.22 shows yields of upland rice with 'complete' fertilizer treatment, 'maintenance' fertilizer treatment, and no fertilization. The declining yields even with 'complete' fertilization were attributed to: (1) insufficient rates of fertilization, (2) poor germination, (3) increased insect attacks, and (4) soil compaction.

One of the steps later taken to remedy declining production, in addition to increasing applications of major nutrient elements, was the application of the micronutrients: zinc, manganese, iron, copper, boron, and molybdenum. These micronutrients substantially improved crop yields (Villachica and Sánchez, 1978).

It is well known that plants require these micronutrients for growth (Frieden, 1972), but vegetation dynamics on the lightly and moderately disturbed sites

Fig. V.21. Three-year-old mangrove vegetation, in 1983, that was planted along tidal rivers near Ho Chi Minh City, Vietnam. The original mangrove forest was completely destroyed by herbicides in the late 1960s.

discussed so far appeared to be controlled by major nutrient elements. Continuous cropping removes both the major nutrient and micronutrient elements and if only the major nutrients are replaced, it is clear that eventually micronutrient losses will become critical.

(d) Fruit plantations

Several examples have shown that sustainable economic yield without either a fallow period or fertilization may be difficult in wet tropical regions. However,

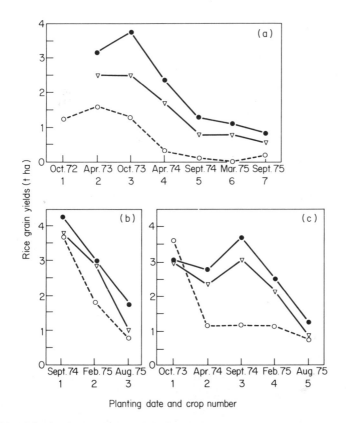

Fig. V.22. Yields of a continuous rice cropping system in eastern Peru as a function of time since clearing: (a) 1972; (b) 1974; (c) 1973. Initially, the 'maintenance' fertilization program consisted of 40 kg of nitrogen, 20 kg of phosphorus and 30 kg of potassium per hectare for each crop, but later nitrogen was increased to 67 kg/ha and potassium to 55 kg/ha. The 'complete' treatments received an initial application of 80 kg of nitrogen, 100 kg of phosphorus, and 80 kg of potassium per hectare and were limited to pH 5.5. Subsequent crops each received 100–26–80 kg/ha of N, P, K. ● — ●, 'Complete'; ▽ — ▽, 'maintenance'; ○ — ○, unfertilized. [Adapted from Villachica, *et al.* (1976).]

systems for continuous production without inorganic fertilizer inputs have been developed in several tropical regions.

The crop species in such systems is usually a shrub or tree with a large total biomass relative to the harvested biomass (Alvim, 1981). Often, only the fruit is harvested. Cacao and coffee are examples. Sustained production without inorganic fertilizer input can be attained only when the crop is grown together with other trees, often nitrogen-fixing legumes. Large leguminous trees increase the availability of nitrogen for coffee plants (Aranguren *et al.*, 1982). The roots of these trees can penetrate deeper and may be able to take up nutrients beyond the reach of the coffee plants (Beer, 1982). These nutrients are returned to the soil in litterfall which, as an input on the soil surface, would be readily available to the coffee trees.

Coffee is grown with a variety of other tree species (Fuentes Flores, 1979). There is a 'rustic' system which uses the natural forest as shade. The plantation is made by clearing out small trees and shrubs on the forest floor to make room for the coffee plants which then grow under near-wild conditions. 'Traditionally' managed coffee plantations combine fruit trees, coffee plants, and leguminous shade trees. This system is often practiced by peasant farmers growing crops on very small plots.

The owners of medium or large farms often cultivate a monocrop of coffee, with shade provided by widely spaced leguminous trees of a single species. Sometimes no shade trees are planted, in which case production may be higher, but sustained yield requires fertilizer input (Bornemisza, 1982). Such a system is economical only as long as fertilizer prices are low and coffee prices are high.

(e) *Pastures*

Pasture has been ranked the least desirable of all possible uses of converted Amazon forest (Goodland, 1980), because of the land degradation that accompanies overgrazing. When too many cattle graze a tract of land, the vegetative cover is reduced, root biomass is decreased, and the soil becomes more susceptible to erosion during heavy rains. Trampling and compaction of the soil by too many cattle prevent the establishment and growth of grass. Water cannot easily infiltrate the soil and, when heavy rains occur, run-off over the bare soil surface creates gullies. Once a gully becomes established, it increases rapidly in size with each succeeding rainstorm (Denevan, 1981). Highly tolerant weeds, many of them poisonous to cattle, are better adapted to these new edaphic conditions and quickly become established. Because Amazon pastureland is often held for speculation rather than production, the landowner may be indifferent to these adverse environmental effects (Hecht, 1981).

In dry areas, overgrazing can lead to desertification because cattle, goats, and sheep destroy the vegetation that holds the sandy soil in place. Approximately 65×10^6 ha of previously productive land in the southern portion of the Sahara are estimated to have become desert during the last 50 years. In the Sudan, the desert is reported to have advanced 100 km in the 17 years prior to 1975 (Novikoff, 1983).

In contrast to these very pessimistic descriptions of the effect of rain forest conversion to pasture, other scientists have claimed highly beneficial effects: 'The use of Amazonian forest areas as cultivated pastures for several years results in an improvement of the soil chemical properties' (Souza Serrão *et al.*, 1978); 'With the formation of pastures, after cutting and burning the forest, there occurs an increase in levels of most nutrients, and these higher levels are maintained for several years of active grazing' (translated from the Portuguese; Falesi, 1976).

Some light has been shed on the controversy by the construction of models showing trends of soil organic matter and nutrients when Amazon forest is

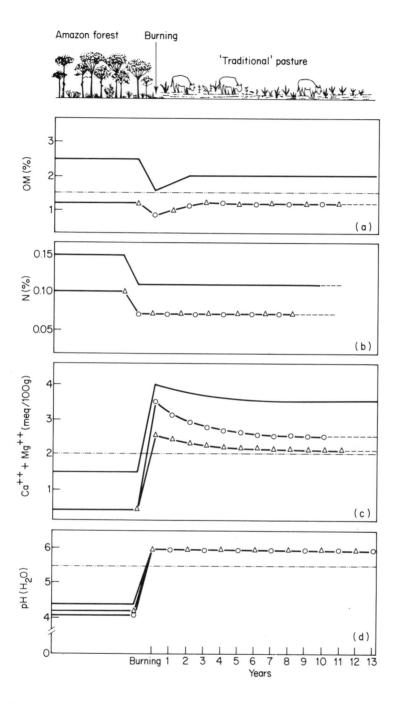

Fig. V.23. Changes in soil organic matter (a), nitrogen (b), calcium plus magnesium (c), and soil pH (d) following conversion of forest to *Panicum maximum* pasture: —, yellow latasol (oxisol), very heavy texture; $\circ - \circ$, red yellow podzolic (ultisol), medium texture; $\triangle - \triangle$, dark red latosol (oxisol), medium texture; –..–..–, critical level (4). [Adapted from Souza Serrão *et al.*, (1978).]

136

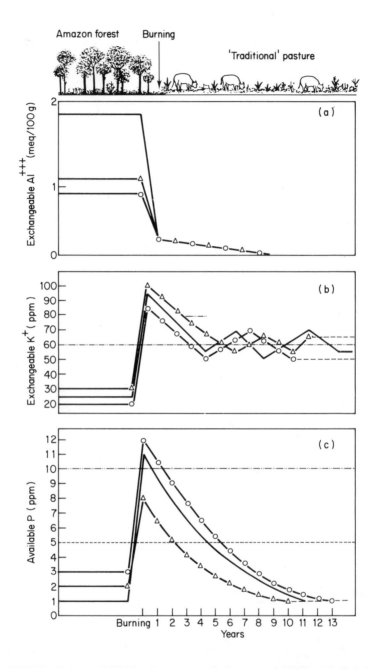

Fig. V.24. Changes in available or exchangeable aluminiun (a), potassium (b), and phosphorus (c) following conversion of forest to *Panicum maximum* pasture: —, yellow latasol (oxisol), very heavy texture; ○ — ○, red yellow podzolic (ultisol), medium texture; △ — △, dark red latosol (oxisol), medium texture; –··–, standard critical level (4) – – – –. proposed critical level. [Adapted from Souza Serrão *et al.* (1978).]

converted to pasture (Souza Serrão *et al.*, 1978). When the forests are cut and burned, both total soil organic matter and nitrogen immediately drop (Fig V.23(a) and (b)). Increase in nitrogen and organic matter following burning depends on the establishment of nitrogen-fixing species in the pasture. Calcium, magnesium (Fig. V.23(c)), and soil pH (Fig. V.23(d)) all rise during the burn and remain well above the levels of the original forest, probably because of repeated burnings. Exchangeable aluminum drops after the burn (Fig. V.24(a)). Following an initial increase, potassium levels fall, but not to the level of the original forest (Fig. V.25(b)). Available phosphorus (Fig. V.24(c)) clearly shows the biggest decline during the course of pasture utilization. Phosphorus is the nutrient which is most critical for sustaining yield in these pastures. According to these models, available phosphorus drops despite maintenance of pH close to 6.0. The decline may be caused by a depletion of total phosphorus stocks.

It is clear that, when the forest is felled, there is a transfer of nutrients from the living biomass to the forest floor. When the biomass is burned, some nitrogen and sulfur are lost, but large quantities of other nutrients are deposited in ash. As the ash is carried down into the soil, there indeed is an increase

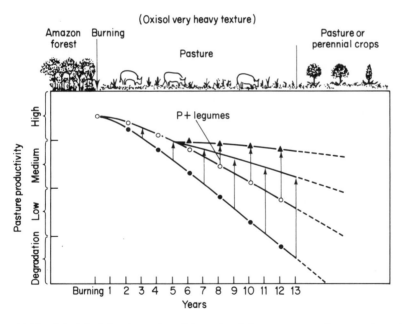

Fig. V.25. Model of the dynamics of the animal–forage grass–soil system in a clayey Amazon oxisol, with and without the addition of phosphorus, and of nitrogen-fixing species: ○ — ○, 'traditional' at optimum grazing pressure; ● — ●, 'traditional' at grazing pressure above optimum; ▲ — ▲, 'improved' at optimum grazing pressure; —, 'improved' at grazing pressure above optimum. 'Improved' pasture means grass and legume and phosphorus; 'traditional' pasture means grass (*P. maximum*). [Adapted from Souza Serrão *et al.* (1978).]

of soil fertility. However, other studies have shown that there are simultaneous nutrient losses from the soil through leaching, volatilization, and fixation (Buschbacher, 1984). When the whole ecosystem is considered, not just the soil, there is a loss of nutrients. A statement that conversion of forest to pasture enriches the soil is misleading, because enrichment of the soil is at the expense of the forest.

The question of soil compaction and soil erosion following conversion of forest to pasture is separate from the question of nutrients. The physical changes in the soil occur as a result of overstocking and overgrazing (Fearnside, 1978). When pasture is lightly grazed, these problems may not occur (Buschbacher, 1984).

In practise, tropical pastures are often burned when productivity declines. The periodic burns raise the soil pH and bring the phosphorus back to an available form. However, each burn results in a smaller increase in phosphorus than the previous burn, because the declining stocks of calcium, potassium, and magnesium result in smaller pH changes. Eventually, nutrient levels become too low to sustain grasses palatable to cattle. Ecosystem dynamics are similar to those described for savannazation.

A model of the response of Amazonian pastures to phosphorus fertilization and to the establishment of nitrogen-fixing legumes is given in Fig. V.25. Clearly, these measures extend the useful life of the pasture. However, in many tropical areas fertilization is not practical. The cost of fertilizer is high. Where there are no roads or railroads, as in much of Amazonia, transport costs can exceed the cost of the fertilizer itself. To this must be added the cost of transporting the livestock to market. Sometimes such costs are not considered when choosing to clear tropical forest for beef production.

8. Disturbances, of moderate intensity, large size, long duration

(a) Pulp plantation

One of the largest man-caused perturbations of tropical forest began in 1967 when billionare industrialist Daniel Ludwig bought 1.6×10^6 ha of virgin Brazilian rain forest to establish a pulpwood plantation (Fearnside and Rankin, 1982). This plantation, known as the Jarí project, has been one of the most controversial forestry projects ever undertaken (Kinkead, 1981; Time, 1976, 1979, 1982). Early criticisms of the project came from ecologists who warned that the infertile soils of the eastern Amazon Basin would limit productivity (Goodland and Irwin, 1975; Fearnside and Rankin, 1980, 1982).

Large-scale plantings of Gmelina arborea commenced in 1969 on sites that were cleared by bulldozers. The slash was moved into long rows (windrows) and burned. The Gmelina is reported to have grown well along the windrows but to have failed in other areas (Greaves, 1979). Forest clearing by heavy machinery was abandoned for three reasons (Posey, 1980): (1) it disturbed the already skimpy topsoil; (2) it compacted the topsoil; and (3) it was expensive.

Heavy equipment was replaced by laborers with axes and chainsaws.

The best growth of *Gmelina* was reported to be about 14 t ha^{-1} a^{-1} (Woessner, 1982). On plots studied intensively by C.E. Russell (1983) biomass accumulation was between 8 and 12 t ha^{-1} a^{-1}. However, the average productivity in a large sample of plots throughout Jarí was only 3–6 t ha^{-1} s^{-1} (Schmidt, 1981). This is considerably less than *Gmelina* production rates in other regions of the world and is also below the average production rates of other pulpwood species (Wadsworth, 1983). *Gmelina* production at Jarí was 40 percent below that originally projected (Kinkead, 1981) and, as a result, trials with other species began.

Many of the new sites were planted with *Pinus caribaea* var. *hondurensis* or with *Eucalyptus deglupta* (Posey, 1980). Many of the poorest *Gmelina* plantations were ripped out and replanted with these other species (Kinkead, 1981). The best growth of pine was 12–16 t ha ^{1} a ^{1} (C.E. Russell, 1983), but as in the case of *Gmelina*, average production throughout the Jarí plantation was lower, being less than 6 t ha ^{1} a ^{1} after seven years (Schmidt, 1981). *Eucalyptus* showed better growth in some cases, but there was considerable variation within sites, and patches of very reduced growth were noticeable (Hartshorn, 1981).

To determine the relationship between soil fertility and plantation productivity at Jarí C.E. Russell (1983) carried out a study of nutrient cycling and net above-ground wood productivity in a series of plantation sites and in the undisturbed primary forest.

The results showed that conversion of primary forest to pine plantation increased annual wood productivity from an average of 12 t ha^{-1} a^{-1} in the primary forest to as much as 16 t ha^{-1} a^{-1} in some of the better pine plantations. However, this increase in productivity occurred at the expense of nutrient stocks which had been stored in the above-ground biomass of the primary forest. In a plot where slash and ash from the primary forest had been cleared away mechanically before planting, diameter growth was 31 percent lower and height growth was 45 percent lower than that in an adjacent stand, where slash and ash had not been cleared away (C.E. Russell, 1983).

The dynamics of four major nutrients during the growth and harvest of the first plantation and during the establishment of the second plantation (projected) are shown in Fig. V.26 (Jordan and Russell, 1983). To the left of time zero, on the x-axis, are stocks of nutrients in the undisturbed primary forest. The ecosystem compartments are stacked vertically, that is, values along the y-axis are cumulative. For example, there are 51 kg of calcium per hectare in the soil of the primary forest. Above that is added the 140 kg of calcium per hectare in the litter, and above that the 1231 kg of calcium per hectare in the vegetation, for a total calcium content of 1422 kg/ha in the entire ecosystem. At the time the primary forest was cut, many of the trunks were removed, some for processing at a sawmill and some for fuel in the pulp factory or in the local power plant. Clearly, the removal of calcium and potassium from the site due to the harvest of logs constituted the most significant loss of these cations from

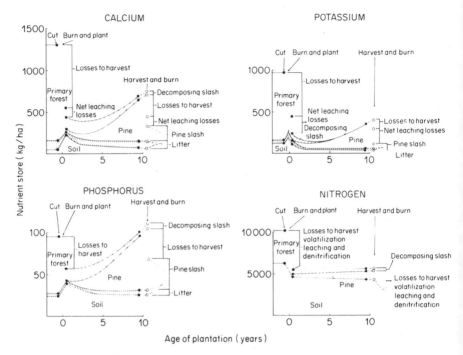

Fig. V.26. Dynamics of four nutrients during the conversion of primary Amazon forest to pine plantation, during the growth and harvest of the first wood crop, and during the establishment of the second crop (projected): ● measured; ○, estimated; – – –, projected. See text for detailed explanation.[Adapted from Jordan and Russell (1983) with permission of Pergamon Press.]

the ecosystem. Net leaching losses were significant only during the first year or two and, thus, are plotted as a single event for simplicity.

The transfer of calcium and potassium from biomass to soil via ash can be seen as a sharp pulse at time zero. Stocks of these nutrients in the soil and in decomposing slash decreased throughout the first rotation. Most of these stocks were transferred to the growing pines. Stocks of calcium and potassium in the pine trees also increased because input from wet and dry atmospheric deposition exceeded leaching losses after the pines became well established at about year three.

In contrast to calcium and potassium, phosphorus stocks appear to be slightly higher in the mature pine plantation than in the primary forest. However, soil phosphorus is plotted as available (acid extractable) phosphorus, which ranges from 6 percent of total phosphorus in surface soil at the study site to 0.3 percent at 1 m depth (C.F. Jordan, unpublished data). Pines and their associated ectomycorrhizae apparently are highly effective at extracting phosphorus from relatively tightly bound forms in the soil. When total phosphorus is considered, stocks change very little during cutting, burning, and growth of the pines. The major change is loss of phosphorus in trees removed from the site.

The pattern of nitrogen dynamics is similar to that in other rain forest disturbances. During the first few years following cutting of the primary forest, nitrogen levels in the soil do not appear low enough to inhibit crop growth. However, nitrogen is restored very little during crop or plantation growth. Nitrogen losses continue to exceed gains, and as stocks decline during the second and subsequent rotations, nitrogen could become critical to tree growth.

It appears that wood production during the second rotation at Jarí will decline, because it has started with a lower stock of soil nutrients than did the first rotation. Because of low productivity, it is not clear how long the project can continue. In 1982 the project was sold at a loss of hundreds of millions of dollars (Fearnside and Rankin, 1982).

(b) Mayan agriculture

The history of the lowland rain forest areas of Guatemala and Mexico bears on the question of the long-term ability of long-disturbed ecosystems to recover. Most of the area in Guatemala which supported the ancient Mayan civilization is now covered by high forest, and much of the Mayan area in Mexico has been covered until recently. Less than 1000 years ago, these areas were intensively cultivated (Hammond, 1982; Wiley, 1982; Turner and Harrison, 1981; Matheny and Gurr, 1979). Although a decrease in soil fertility was probably only one of several factors that contributed to the decline of the Mayan civilization, the relevant point is that the existence of high forest today demonstrates its ability to regenerate even after large-scale, long-term disturbance.

9. High intensity disturbance: *removal of soil*

In disturbances of high intensity, not only is the structure of the ecosystem destroyed, but the soil is also severely affected. In the tropics, natural disturbances of high intensity may be more frequent than commonly realized. For example, in 1976 two earthquakes struck near the southeastern coast of Panama. Landslides associated with the tremors denuded about 54 km^2 of steep terrain originally covered by tropical rain forest. Earthquakes are frequent in New Guinea and, in 1935, denuded 130 km^2 of tropical forest (Garwood *et al.*, 1979).

Earthquakes are not the only cause of major landslides. Extremely heavy rainfall in Puerto Rico in 1979 during hurricane 'David' (Jordan and Farnworth, 1982; Lugo *et al.*, 1983) saturated the subsoil on a slope in the Luquillo forest, causing a major landslide. Five years previously, Lewis (1974) had predicted that landslides were likely in Luquillo forest, especially during heavy rains. The prediction was based on the observation that tree roots were

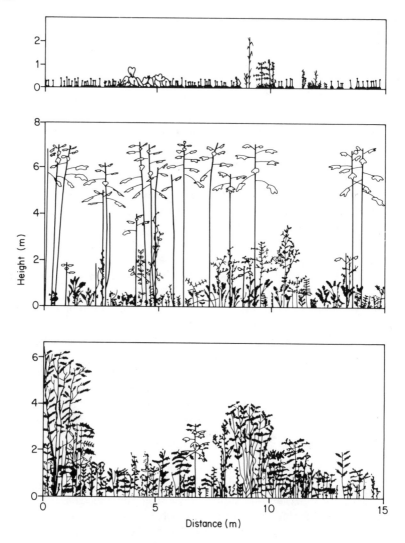

Fig. V.27. Profile diagrams representing the vegetation present on a 1m × 15 m transect three years after caatinga forest was cut (bottom), cut and burned (middle), and cut and bulldozed (top). [Adapted from Uhl *et al.* (1982b) with permission of *Oikos*.]

concentrated near the soil surface, and that there were few roots in the subsoil to prevent soil-creep.

The time required for such denuded areas to recover depends on the physical and chemical nature of the exposed substrate. If there is a layer of particles fine enough to hold at least a small amount of water and nutrients, vegetation re-establishes quickly. For example, on Eniwetok atol in the southwestern Pacific, nuclear testing was carried out through 1958. By 1964, several species of trees up to 4 inches (10 cm) in diameter were growing within 15 feet (5 m) of crater lips (Koranda, 1965). However, in the middle of open-pit phosphate mines on Nauru Island in the southern Pacific, it may take centuries for forest to re-establish, even in a modified form (Manner *et al.*, 1984).

Uhl *et al.* (1982b) compared revegetation on rain forest sites that had been cut and abandoned; cut, burned, and abandoned; and cut and bulldozed. The removal of topsoil on the bulldozed site had a major influence on the species composition and productivity of the colonizing plant community (Fig. V.27). A similar experiment in Nigeria showed productivity of corn differed greatly, depending on the nature of the remaining soil material (Lal, 1981a).

It does not necessarily take long for sufficient nutrients to accumulate on a denuded site to permit the re-establishment of woody vegetation. Organic matter of any kind, such as leaf litter or bird droppings, carried into a denuded area can serve as a substrate for tree growth. For example, Ohi'a (*Metrosideros collina*), a secondary successional tree which grows on the slopes of Hawaiian volcanos, rapidly establishes in pockets of organic matter which accumulate in crevices in solid lava flows (C. F. Jordan, personal observation).

The slowest vegetative recovery occurs when disturbance results in the exposure of solid rock, when there is little organic matter to form a substrate, and when a seed source is distant. An example is the formation of a new island following the complete volcanic destruction of an oceanic island such as Krakatoa (Krakatau), west of Java, in 1883 (Francis and Self, 1983). The patterns of primary succession on such volcanic 'deserts' were studied by Hirose and Tateno (1984), who showed that the rates of nitrogen accumulation, and the nature of nitrogen compounds, were important factors governing revegetation.

10. A model of rain forest dynamics during disturbance

The discussion of changes in tropical forest ecosystems during various types of disturbance has revealed several patterns of nutrient cycling and vegetation dynamics in different regions and under different circumstances. For example, there are many similarities between the conversions of Amazon forest to pasture, African forest to savanna, and Asian forest to bamboo brake.

Although nutrient and vegetation dynamics differ quantitatively between sites, the qualitative similarities can be synthesized into a general summary model (Fig. V.28) which describes structural and functional dynamics during disturbance of tropical forests.

There is an almost continual input of nutrients into the ecosystem from 'external' sources (a). These inputs come primarily from precipitation, dry-fall, nitrogen fixation, and also from mineral weathering where the parent material is relatively fertile. There is a continual loss of nutrients (d) due to leaching, erosion, and denitrification. In the closed forest (b), inputs are closely balanced by outputs. However, the forest is never in steady state for more than a short time, because its equilibrium is continually interrupted by tree fall (gap dynamics, (c)) or by larger disturbances, such as high winds. These events

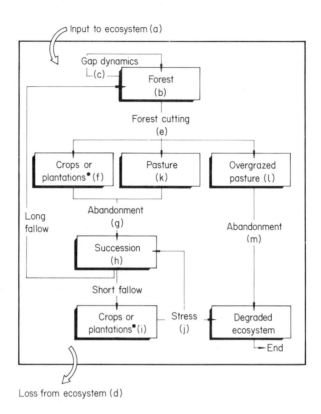

Fig. V.28. Summary model of ecosystem dynamics during disturbances of wet and moist tropical forests: ●, soil high in organic matter; ■, soil low in organic matter.

release nutrients stored in large, immobile compartments such as wood, making them available for plant growth. Total stocks of nutrients in the forest change very little as a result of such natural disturbances.

If the forest is cut (e), some nitrogen and sulfur are lost through volatilization. The relatively large biomass stores of calcium, potassium, and magnesium are transferred to the soil, where they are available for plant uptake. They also raise the pH of the surface soil, thereby making phosphorus more readily available. Nutrients in unburned slash and soil organic matter also become available through decomposition.

The cleared forest is frequently used for annual cropping, fruit tree or pulpwood plantations, or pasture ((f), (k)). The initial rate of production may be good if the original forest had high biomass and had accumulated a large store of nutrients. Further, soil structure may be good because pores from decaying tree roots provide ample aeration and growth space for the crop species. In addition, crop pests have not had an opportunity to build up.

During cultivation, nutrients are lost from the site through harvest, leaching, volatilization, and fixation. As nutrients are lost or become unavailable, crop productivity decreases. Other factors, such as insect pests, can also play a role in declining productivity.

If a disturbed area is fallowed after a relatively short period of exploitation (g), the fallow vegetation (h) is able to restore the fertility of the soil. As the successional vegetation grows, it incorporates nutrients from both the atmosphere and weathering minerals. This is the mechanism by which the site regains fertility. Any subsequent crop rotation derives nutrients from the decomposing remains of the fallow vegetation. In regions with highly weathered mineral substrates, such as the central and eastern Amazon Basin, rebuilding of nutrient stocks occur almost entirely through atmospheric deposition (Jordan, 1982a) and the process may take a long time. In contrast, where mineral substrates are relatively little weathered (e.g. volcanic material), replenishment of nutrient stocks by fallow vegetation may be rapid, because nutrient stocks have not disappeared, but rather have become unavailable (Sanchez, 1976).

If a site is continually disturbed, such as annually burned pasture land (l), productivity declines. One reason that pastures on acid soils are burned is to raise soil pH and, thus, improve phosphorus availability. Burning also eliminates woody vegetation. However, continual burning depletes soil organic matter and nitrogen stocks. Burning also converts calcium and potassium to soluble forms and, eventually, leaching losses of these nutrients become important.

The same sort of degradation can occur when fallow cycles are too short ((i), (j)). Nutrient stocks are lowered to the point where forest vegetation cannot re-establish. Whether such degradation can be permanent in the tropics remains an open question.

D. Chapter summary

Chapter I developed evidence that the potential for nutrient loss in the wet tropics is high. However, subsequent chapters indicated that actual nutrient loss in undisturbed tropical forests is usually low, because the native vegetation has adapted to the high leaching potential through a variety of nutrient conserving mechanisms. An important point is that these mechanisms are effective only so long as the native forest is undisturbed.

This chapter examined the changes in nutrient dynamics in humid tropical ecosystems as a result of disturbance. Change was discussed as a function of disturbance intensity, size, and duration. Changes ranged from virtually undetectable (in small, short, non-intense disturbances such as natural tree falls) to almost total (in areas denuded by landslides, or volcanic activity). Most of the disturbances caused by man fall somewhere in between the two extremes.

There was strong evidence that nutrient loss or fixation of phosphorus is an important cause of the declining productivity experienced in most tropical agriculture. The application of fertilizers often can reverse declining productivity, but the economics of fertilization appear prohibitive in relatively underdeveloped and remote regions. Agricultural and forestry systems that maintain or restore soil fertility through the conservation or replenishment of soil organic matter are more promising.

Chapter VI

Conclusion

A. Paradox of forests in the humid tropics

The objectives of this book have been to present the evidence that nutrients are the factor most often limiting production in the humid tropics and to discuss the implications of nutrient scarcity in both disturbed and undisturbed tropical ecosystems.

Clearly, there are site to site differences within the tropics, and even within regions. The intent of this book has not been to predict nutrient scarcity at any particular site or particular time. Rather, the intent has been to consider nutrient deficiency as a generic problem, much as water scarcity is a generic factor limiting productivity in desert regions. In deserts, plant productivity at a particular site and time may be limited by hard-pan, wind, animals, or other factors. Nevertheless, to cultivate the desert, one has to deal first and foremost with the problem of water. Similarly, to cultivate tropical forest lands, agriculturalists, businessmen, planners, politicians, and others must deal first and foremost with the problem of nutrients.

Experienced tropical ecologists, foresters, and agronomists might ask why this point needs to be elaborated. To them, the nutrient problem is obvious. But to many others, the point is not at all obvious. Even today, many scientists and laymen believe that tropical rain forests have a virtually unlimited potential as the future bread basket of the world. For example, Thurston (1969) has stated, 'The most realistic short-range method of solving the world's food crisis is to increase food production through conventional agriculture in the tropics.'

Why does this belief in high potential productivity in the humid tropics persist so strongly, despite abundant evidence to the contrary? The question has been addressed with reference to the forests of the neotropics, as follows:

The first European explorers of the American tropics were the sixteenth-century conquistadors who came in search of gold. For them, the rain forests of Central and South America were obstacles that interfered with the quest for riches. The great size, density, and lushness of the forests impeded them, and they blamed failed expeditions on the forests' impenetrability. For example, in a chronicle of the

147

travels of Hernando Cortez through the lowland rain forest of Mexico, Diaz del Castillo wrote in the early sixteenth century (1916, p.14): 'The forest was so excessively high and thick we could seldom see the sky, and, when [the soldiers] attempted to climb some of the trees in order to survey the country, they could see nothing at all, so dense was the forest, and two of the guides we had brought with us fled, and the one who remained was so ill that he could explain nothing about the road or any other matter.'

It was three centuries later that another type of explorer, the explorer-scientist, traveled to the rain forests of the New World. These scientists, among whom were Humboldt, Bates, and Wallace, viewed the forest differently than did the conquistadors. The explorer-scientists looked at the forests as a potential source of economically important products. And although they sometimes encountered valuable species of plants, the qualities of the forests that impressed the scientist-explorers most were those that had most discouraged the conquistadors – that is, the forests' size, density, and apparent vigor. For example, Wallace (1878, p. 65) apparently believed that the Amazon forests had great productive potential when he wrote: 'The primeval forests of the equatorial zone are grand and overwhelming by their vastness and by the display of a force of development and vigour of growth rarely or never witnessed in temperate climates.'

These scientists knew that in Europe the size of the trees in a forest generally indicated the potential for agricultural use of the soil supporting the forest. Large trees usually meant the soil would be highly productive. Small trees usually meant that crop yields from that soil would be small. The association between large trees and high agricultural productivity resulted in the belief that the Amazon Basin had extremely high potential for timber and crop yields.

This belief stimulated many attempts to exploit the area commercially. Sioli (1973) chronicled the efforts at colonizing and developing the Amazon Basin for agriculture and forestry. The most important point of Sioli's account is that all large-scale developments have failed, often because of the nutrient poverty of Amazon soils. Although some individual farmers have been able to carry out subsistence agriculture (Moran, 1981) or cultivation of certain perennial crops (Alvim, 1978), large-scale efforts at production of cash crops for export have not been successful. Even more striking than the failures is the general refusal to believe that the poor soils are at fault. Failures are blamed on poor administration, lack of adequate labor force, and insufficient official support, but rarely, if ever, on the productive potential of the soils (Sioli, 1973). [Reprinted from Jordan, 1982c, by permission of Sigma Xi].

B. Principles of management for humid, tropical ecosystems

Despite the frequent problem of nutrient scarcity in the humid tropics, there are numerous examples of successful tropical agriculture. Plantations of rice, rubber, cacao, and bananas (Alvim, 1978) are only a few examples. Often, these economically valuable crops are grown in relatively rich alluvial or volcanic soils. However, even on such soils, fertilization is usually necessary.

Management techniques, such as agroforestry, which minimize fertilizer requirements for sustained yield were mentioned in Chapter V. In areas where commercial fertilization is not economical, such techniques seem to hold the most promise for sustaining productivity. Details of the various management schemes are beyond the scope of this book, but references to some recent publications have been given (Chapter V). Despite the great variability of such techniques from site to site, and from crop to crop, there are several principles which the techniques share, regardless of the site or crop. Those principles are: maintenance of structural diversity, maintenance of soil organic matter, minimization of soil disturbance, and control of the size and shape of the disturbed area.

1. Structural diversity and improved resource utilization

Structurally complex and diverse plant communities may have advantages over simple monocultures in the humid tropics. There are a number of reasons for the advantages (Hart, 1980; National Research Council, 1982).

A diverse community is likely to have a well-developed, complex root system that extends well below the soil surface, impedes erosion and nutrient loss, and exploits more efficiently the stocks of nutrients distributed throughout the entire soil horizon. Complex communities benefit from the presence of large trees which enrich the topsoil by sending down deep tap roots that can exploit nutrients released by the weathering of parent rock and redeposit them as leaf litter on the soil surface (Beer, 1982). In a complex community, mutualistic interactions may occur between species, enhancing the production of the ecosystem as a whole. Nitrogen-fixing plants, which can benefit their associates, are a well known example (National Research Council, 1982). The presence of many species also ensures that the soil is at least partially covered all the time, since harvest of different species usually does not occur simultaneously.

Diversity inhibits the exponential groth of herbivore populations, because host plants are spatially separated and because a diverse habitat can support a greater variety of predators (Karel et al., 1982). A diverse habitat is also more likely to include species which produce allelochemicals effective against herbivores and weeds (Gliessman et al., 1981).

A structurally diverse community can utilize a greater proportion of incoming solar radiation due to its complex spacing of leaves, thereby producing higher yields (May and Misangu, 1982). Denser canopies also lessen the impact of high-intensity rains on the soil.

These advantages may account, in part, for the phenomenon called 'overyielding', wherein the yield of a polyculture exceeds that of any of its component crops when they are grown as monocultures (Vandermeer, 1981; Vandermeer et al., 1983).

Not all these advantages occur in all structurally diverse agro-ecosystems. For example, Brown (1982) and Ewel et al. (1982) tested these principles in a series of tropical communities ranging from a simple maize monoculture through a complex wooded garden. At their site, they found that net primary productivity was not related to ecosystem complexity. They also found that leaf damage from insects was remarkably similar in all plots, even though there were substantial species-to-species differences. Herbivores seemed to consume a nearly constant proportion of the total amount of leaves present. They also found that a sweet potato monoculture was as effective in intercepting solar radiation and reducing the impact of rainfall as more diverse ecosystems, but a maize monoculture was less effective. The big advantage of structural diversity of their sites was that exploitation of the soil reservoir was much more complete in the complex ecosystems than in the simple ones (Fig. V1.1). Fine roots and total roots were much more abundant in the complex ecosystems. The quantity of fine roots is probably a good indicator of an ecosystem's potential for nutrient uptake, and the presence of all size classes of roots indicates its ability to resist erosion.

Because Ewel *et al.* (1982) could not document other advantages of structurally diverse ecosystems at their study site does not mean that they never occur. For example, Risch (1981) found that beetle populations were lower in polycultures than in similarly treated monocultures. An entire volume on intercropping (Keswani and Ndunguru, 1982) describes many experiments in complex agro-ecosystems, some of which demonstrated advantages of complexity, and others which did not.

Another advantage of mixed cropping systems is economic. Crop failure can be devastating when there is only a single crop species, but the impact of failure of one species in a polyculture is less since the failure is offset by other crops.

2. Maintenance of soil organic matter

Although there are many different mechanisms which enable fallow vegetation to take up nutrients, prevent erosion, and improve the structure of degraded soil, all the mechanisms act to increase the amount of soil organic matter.

In mixed cropping systems, leaf litter and fine roots shed by perennial species provide a continual source of nutrients for the fast-growing annual crop plants. Many annual crop plants bred for high productivity cannot obtain nutrients from degraded soil, but can do so if the soil is enriched with decomposing organic matter.

The most desirable cultivation methods are those which leave the below ground ecosystem intact, or which rapidly redevelop their own large below ground ecosystem. Such a system consists of a large pool of soil organic matter, which serves as a nutrient storage reservoir, and an active and diverse microbial community. The microorganisms are extremely important in preventing nutrient loss, and in supplying nutrients to crops, because the nutrients they incorporate are slowly but continuously released in soluble form.

Soil organic matter is more effective than many commercial inorganic fertilizers, which are highly soluble and rapidly leach away before they can be taken up by crop plants. In the case of phosphorus, rapid release results in rapid binding by the iron and aluminum. 'Slow release' fertilizers have been developed to overcome this problem and rock phosphate sometimes is effective (Olson and Engelstad, 1972), but soil organic matter is a naturally occurring slow-release fertilizer. Mulching and green manure have been tested and both appear to hold promise as replacements for inorganic fertilizers in intensive cropping in the humid tropics (Wade and Sánchez, 1983).

Organic matter improves the structure of some soils by aggregating clay

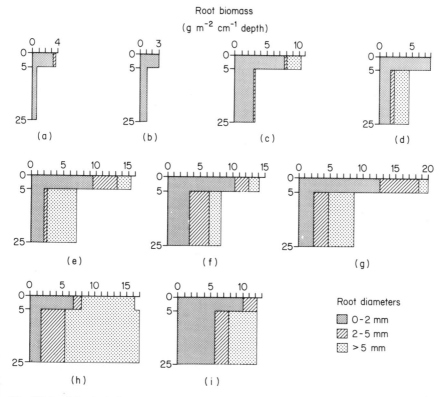

Root biomass
(g m^{-2} cm^{-1} depth)

Root diameters
☐ 0-2 mm
▨ 2-5 mm
▨ > 5 mm

Fig. VI.1. Vertical distribution of root biomass by size class in nine tropical agricultural and successional ecosystems. Communities arranged in order of increasing complexity from left to right and top to bottom: (a) young maize; (b) sweet potato; (c) mature maize; (d) gmelina; (e) succession; (f) shaded coffee; (g) cacao–plantain–cordia; (h) succession mimic; (i) wooded garden. [Adapted from Ewel *et al.* (1982) with permission of Elsevier Science Publishers.]

particles (Sánchez, 1976). Soil that forms such blocks or crumbs drains rapidly and is readily penetrated by fine roots. Decomposition *in situ* of entire roots results in channels which are very effective in carrying off water rapidly during intense tropical storms (A.E. Russell, 1983).

3. Minimizing disturbance of the soil

One important reason why minimum tillage agriculture is effective is that it minimizes disturbance of the soil structure. It causes less water runoff and soil erosion than do mechanical clearing and conventional tillage (Table VI.1). Consequently, nutrient loss is lower when manual clearing and no-till agriculture are used (Table VI.2). Site preparation with the use of herbicides rather than shearing and disking to control sprouting, also holds promise for nutrient conservation (Vitousek and Matson, 1984). Herbicides do not disturb the soil structure and, consequently, their use conserves soil organic matter and nutrients.

Table VI.1

Effects of deforestation methods and tillage systems on sediment density, water runoff, and soil erosion from a maize–cassava rotation in a lowland forest area of Nigeria* (Lal, 1981b)

Clearing treatment	Tillage system	Sediment density (g/l)	Water runoff (mm/year)	Soil erosion (t ha^{-1} a^{-1})
Traditional clearing	Traditional seeding	0.0	2.6	0.01
Manual clearing†	No-tillage	3.4	15.5	0.4
Manual clearing†	Conventional tillage	8.6	54.3	4.6
Crawler tractor/shear blade	No-tillage	5.7	85.7	3.8
Crawler tractor/tree pusher	No-tillage	5.6	153.1	15.4
Crawler tractor/tree-pusher	Conventional tillage	13.0	250.3	19.6

* Sediment density reported here was from a rainstorm monitored on May 31, 1979.
† Includes use of chain saws.

Table VI.2

Runoff and nutrient loss in runoff water from different land clearing and soil management treatments in the months of July, August, and September in a lowland rain forest of Nigeria (Kang and Lal, 1981)

Clearing treatment	Tillage systems	Runoff (mm)	Nutrient loss in runoff water (kg/ha)										
			NH_4–N	NO_3–N	PO_4–P	K	Ca	Mg	Na	Fe	Mn	Zn	Total
1. Manual clearing*	No-tillage	5.3	0.03	0.04	tr	1.0	0.4	0.09	1.1	0.2	0.05	0.11	2.8
2. Manual clearing	Conventional tillage	24.1	0.1	0.23	tr	2.4	1.6	0.41	4.3	0.8	0.07	0.09	9.9
3. Crawler tractor/ shear-blade	No-tillage	31.2	0.4	0.7	0.01	3.7	1.4	0.52	3.3	2.1	0.26	0.17	12.6
4. Crawler tractor/ tree-pusher	No-tillage	66.7	0.6	0.5	0.08	6.7	4.0	1.36	7.0	3.1	0.17	0.18	23.7
5. Crawler tractor/ tree-pusher	Conventional tillage	94.6	0.4	1.7	0.07	10.0	7.9	1.30	15.2	4.3	0.35	0.58	41.9
6. Traditional clearing	Traditional seeding	12.8	0.1	0.1	tr	1.1	2.1	0.46	1.1	0.1	0.04	0.09	5.3

tr = trace (<0.01).
* Manual clearing included use of chain saws, while traditional clearing did not.

154

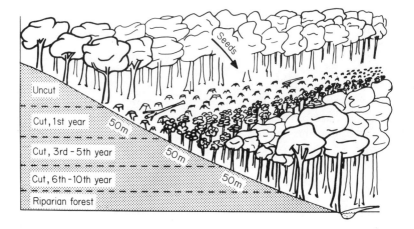

Fig. VI.2. Strip cutting scheme designed to minimize soil erosion and nutrient loss and to maximize the input of seeds and mycorrhizal spores into the disturbed area. [Adapted from Jordan (1982c) with permission of Sigma Xi.]

4. Size and shape of the disturbed area

Another factor important in tropical land use is the size and shape of the cleared area. In areas where small cut areas are interspersed with forest or other dense cover, or where the forest is cut in strips (Fig. VI.2), erosion and nutrient leaching are reduced. One reason is that the uncut vegetation breaks the erosive force of surface water flow. Moreover, the uncut vegetation can take up nutrients in leachate water, thus lessening nutrient loss from the entire area. The practise of leaving forest strips along waterways in areas where the uplands are logged or cultivated is an important soil and nutrient conservation measure (Lowrance *et al.*, 1984; Peterjohn and Correll, 1984). Cutting tropical forests in strips, or 'corridors', has been tried in Africa, but abandoned because of short-term economic problems (Kellog, 1963). Strip cutting has been found to reduce erosion and nutrient loss in temperate forests (Hornbeck *et al.*, 1975), and it may be beneficial in tropical forests where long-term economic and environmental considerations are critical (Jordan 1982c).

Another reason why small clearings are more desirable involves the re-establishment of forest vegetation after disturbance ceases. Birds and mammals, which are often important dispersers of seeds and mycorrhizal propagules (Gómez-Pompa *et al.*, 1972, Mosse *et al.*, 1981, Janson, 1983) will enter and cross small clearings, such as those resulting from shifting cultivation. They are less likely to carry seeds to the middle of a clearing several kilometers in diameter.

5. Economic systems for the humid tropics

Greenland (1975), among others, has pointed out that the reason the much-heralded 'Green Revolution' in the tropics has lost momentum is that high yield-fertilizers, pesticides, tillage and harvesting machinery, proper irrigation, and access to markets is not available to the small farmer, nor is it adapted to his level of education or his normal scale of operations. Furthermore, the small farmer usually has neither capital nor the access to credit needed for this type of agriculture. Even large-scale operations with abundant capital (e.g. the Jarí plantation, Chap. V) have not successfully confronted the high costs of overcoming the environmental constraints of the humid tropics.

An agricultural system for the humid tropics, to be ecologically and economically successful, should include (Greenland, 1975): (1) zero tillage and extensive use of plant residue mulches; (2) mixed crops of high-yielding varieties that are disease and pest resistant; (3) fertilizers to replace the phosphorus and possibly other nutrients removed in produce sold 'off the farm'; (4) legumes, with highly active nitrogen-fixing rhizobia to supply nitrogen to other crops; (5) control of soil pH with ash applications, mulches of deep-rooted species, or lime where it is readily available.

C. Summary of management recommendations

Successful management of ecosystems in the humid tropics incorporates three important principles:

(1) There is a limit to the intensity, size, and duration of disturbance which an ecosystem can withstand and still maintain its productivity and ability to recover.
(2) Native species are often most effective at maintaining the productive capacity of a site or of restoring soil fertility, because they are naturally adapted to local conditions.
(3) Nutrients are very likely to be critical.

References

Alexander, E. B., and J. Pichott, 1979. Soil organic matter in relation to altitude in equatorial Colombia. *Turrialba*, **29**: 183–188.

Alvim, P. T. 1978. Perspectivas de producão agricola na regiâo Amazonica. *Interciencia*, **3**: 243–251.

Alvim, P. T. 1981. A perspective appraisal of perennial crops in the Amazon Basin. *Interciencia*, **6**: 139–145.

Anderson, A. B. 1981. White-sand vegetation of Brazilian Amazonia. *Biotropica*, **13**: 199–210.

Anderson, J. M., J. Proctor, and H. W. Vallack. 1983. Ecological studies in four contrasting lowland rain forests in Gunung Mulu National Park, Sarawak. III. Decomposition processes and nutrient losses from leaf litter. *Journal of Ecology*, **71**: 503–527.

Aranguren, J., G. Escalante, and R. Herrera. 1982. Nitrogen cycle of tropical perennial crops under shade trees. *Plant and Soil*, **67**: 247–258.

Arkley, R. J. 1963. Calculation of carbonate and water movement in soil from climatic data. *Soil Science*, **92**: 239–248.

Ashton, P. C. 1984. Long-term changes in dense and open inland forests following herbicidal attack. In A. H. Westing (ed.) *Herbicides in war, the long-term ecological and human consequences*, pp. 33–37. Taylor and Francis, London.

Atsatt, P. R., and D. J. O'Dowd. 1976. Plant defence guilds. *Science*, **193**: 24–29.

Aubert, G., and R. Tavernier. 1972. Soil survey. In *Soils of the Humid Tropics*, pp. 17–44. National Academy of Science, Washington, D. C.

Auerbach, S. I. 1971. Contributions of radioecology to AEC mission programs. In D. J. Nelson (ed.), *Radionuclides in ecosystems*, Proceedings of the Third National Symposium on Radioecology, pp. 3–8. US Atomic Energy Commission. Washington, D.C.

Aweto, A. O., and O. Areola. 1979. Soil–plant interrelations during secondary succession in the forest zone of Nigeria. In D. U. U. Okali (ed.), *The Nigerian rainforest ecosystem*, Proceedings of the Man and the Biosphere Workshop on the Nigerian Rainforest Ecosystem, January 1979, pp. 243–261. Department of Forest Resources Management, University of Ibadan, Ibadan, Nigeria.

Baker, D. E. 1976. Soil chemical constraints in tailoring plants to fit problem soils. 1. Acid soils. In M. J. Wright, (ed.), *Plant adaptation to mineral stress in problem soils*, Proceedings of a workshop, Beltsville, Maryland, November 1976, pp. 127–140. Agency for International Development, Washington, D. C.

Barnes, R. F. W. 1983. Effects of elephant browsing on woodlands in a Tanzanian national park: measurements, models, and management. *Journal of Applied Ecology*, **20**: 521–540.

Bartholomew, W. V., J. Meyer, and H. Laudelout. 1953. Mineral nutrient immobilization under forest and grass fallow in the Yangambi (Belgian Congo) region. Serie Scientifique No. 57, Publications de L'Institut National Pour L'Etude Agronomique du Congo Belge.

Baur, G. N. 1964. *The ecological basis of rainforest management.* Forestry Commission, New South Wales.

Bazzaz, F. A., and S. T A. Pickett. 1980. Physiological ecology of tropical succession: a comparative review. *Annual Review of Ecology and Systematics,* **11**: 287–310.

Beadle, N. C. W. 1962. Soil phosphate and the delimitation of plant communities in Eastern Australia II. *Ecology,* **43**: 281–288.

Beadle, N. C. W. 1966. Soil phosphate and its role in molding segments of the Australian flora and vegetation, with special reference to xeromorphy and sclerophylly. *Ecology,* **47**: 992–1007.

Beard, J. S. 1944. Climax vegetation in tropical America. *Ecology,* **25**: 127–158.

Beard, J. S. 1949. *The natural vegetation of the Windward and Leeward Islands.* Clarendon Press, Oxford.

Beer, J. 1982. Possible advantages and disadvantages of including shade trees with perennial crops. *Agro-forestry (Turrialba),* August 1982: 8–11.

Benemann, J. R. 1973. Nitrogen fixation in termites. *Science,* **181**: 164–165.

Bennett, C., G. Budowski, H. Daugherty, L. Harris *et al.* 1974. Interaction of man and tropical environments. In E. G. Farnworth and F. B. Golley (eds), *Fragile ecosystems,* pp. 139–182. Spinger-Verlag, New York.

Berish, C. W. 1982. Root biomass and surface area in three successional tropical forests. *Canadian Journal of Forest Research,* **12**: 699–704.

Bernhard-Reversat, F. 1975. Les cycles biogeochemiques des macroelements. In G. Lemee (ed.), *Recerches sur l'ecosysteme de la foret subequatoriale de basse Cote-D'Ivoire: les cycles des macroelements,* Chap. 5. (*La Terre et la Vie,* **29**: 229–254.)

Binkley, D., J. P. Kimmins, and M. C. Feller. 1982. Water chemistry profiles in an early and mid-successional forest in coastal British Columbia. *Canadian Journal of Forest Research,* **12**: 240–248.

Bolin, B., and E. Arrhenius. 1977. Nitrogen – an essential life factor and a growing environmental hazard. *Ambio,* **6**: 96–105.

Borie, F., and H. Zunino. 1983. Organic matter–phosphorus associations as a sink in P-fixation processes in allophanic soils of Chile. *Soil Biology and Biochemistry,* **15**: 599–603.

Bormann, F. H., G. E. Likens, D. W. Fisher, and R. S. Pierce. 1968. Nutrient loss accelerated by clear-cutting of a forest ecosystem. *Science,* **159**: 882–884.

Bormann, F. H., G. E. Likens, and J.M. Melillo. 1977. Nitrogen budget for an aggrading northern hardwood forest ecosystem. *Science,* **196**: 981–983.

Bornemisza, E. 1982. Nitrogen cycling in coffee plantations. *Plant and Soil,* **67**: 241–246.

Bourgeois, W. W., D. W. Cole, H. Reikerk, and S. P, Gessell. 1972. Geology and soils of comparative ecosystem study areas, Costa Rica. Contribution No. 11, Institute of Forest Products, College of Forest Resources, University of Washington.

Bowen, H. J. M. 1979. *Environmental chemistry of the elements.* Academic Press, London.

Boyer, J. 1973. Cycles de la matière organique et des éléments minéraux dans une cacaoyère camerounaise. *Cafe, Cacao, Thé* **17**: 3–23 [cited in Bernhard-Reversat, 1975].

Brady, N. C. 1974. *The nature and properties of soils.* Macmillan, New York.

Brams, E. 1973. Soil organic matter and phosphorus relationships under tropical forests. *Plant and Soil,* **39**: 465–468.

Brasell, H. M., and D. F. Sinclair. 1983. Elements returned to forest floor in two rainforest and three plantation plots in tropical Australia. *Journal of Ecology,* **71**: 367–378.

Brinkman, W. L. F., and J. C. Nascimento. 1973. The effet of slash and burn agriculture on plant nutrients in the tertiary region of Central Amazonia. *Acta Amazonica,* **3**: 55–61.

Brown, B. 1982. Productivity and herbivory in high and low diversity tropical succesional ecosystems in Costa Rica. PhD dissertation, University of Florida, Gainesville.

Brown, N. R. 1981. Problems related to management of pine forests in Central America. In M. Chavarria (ed.), *Simposio international sobre las ciencias forestales y su contribucion al desarrollo de la America tropical, San Jose, Costa Rica, Octubre 1979*, pp. 61–69. CATIA, Turrialba, Costa Rica.

Brown, S., and A. E. Lugo. 1984. Biomass of tropical forests: a new estimate based on forest volumes. *Science*, **233**: 1290–1293.

Brown, S., A. E. Lugo, S. Silander, and L. Liegel. 1983. Research history and opportunities in the Luquillo experimental forest. Southern Forest Experiment Station General Technical Report SO–44, Institute of Tropical Forestry, Rio Piedras, Puerto Rico.

Bruijnzeel, L. A. 1982. Hydrological and biogeochemical aspects of man-made forests in south-central Java, Indonesia. Thesis: Final Report, Vol. 9, Nuffic Serayu Valley Project, ITC/GUA/VU/1, the Hague, Netherlands.

Brunig, E. F. 1974. *Ecological studies in the kerangas forests of Sarawak and Brunei*. Borneo Literature Bureau, Malaysia.

Budowski, G. 1956. Tropical savannas: a sequence of forest felling and repeated burnings. *Turrialba*, **6**: 22–23.

Buschbacher, R. J. 1984. Productivity and nutrient cycling following conversion of Amazon forest to pasture. PhD dissertation, University of Georgia, Athens, Ga.

Carlisle, A., A. H. F. Brown, and E. J. White. 1966. The organic matter and nutrient elements in the precipitation beneath a sessile oak (*Quercus petraea*) canopy. *Journal of Ecology*, **54**: 87–98.

Carlson, E. A. 1983. International symposium on herbicides in the Vietnam war: an appraisal. *BioScience*, **33**: 507–512.

Chandler, T., and D. Spurgeon (eds). 1979. *Proceedings of conference on international cooperation in agroforestry, Nairobi, Kenya, July 1979*. International Council for Research in Agroforestry, Nairobi, Kenya.

Chapin, F. S. 1980. The mineral nutrition of wild plants. *Annual Review of Ecology and Systematics*, **11**: 233–260.

Chapin, F. S. 1983. Patterns of nutrient absorption and use by plants from natural and man-modified environments. In H. A. Mooney and M. Godron (eds), *Disturbance and ecosystems: components of response*. Springer-Verlag, Berlin.

Chapin, F. S., and K. Van Cleve. 1981. Plant nutrient absorption and retention under differing fire regimes. In *Fire regimes and ecosystem properties*, pp. 301–321. US Department of Agriculture Technical Report Number WO-26, U.S. Forest Service.

Charley, J. L., and B. N. Richards. 1983. Nutrient allocation in plant communities: mineral cycling in terrestrial ecosystems. In O. L. Lange, P. S. Nobel, C. B. Osmond, and H. Ziegler (eds), *Physiological plant ecology*, Vol. IV, *Ecosystem processes: mineral cycling, productivity and man's influence*, pp. 5–45. Springer-Verlag, Berlin.

Charley, J. L., and J. W. McGarity. 1978. Soil fertility problems in development of annual cropping on swiddened lowland terrain in northern Thailand. In P. Kundstadter, E. C. Chapman, and S. S. Sabhasri (eds), *Farmers in the forest*, pp. 236–254. University of Hawaii, Honolulu.

Cole, D. W., S. P. Gessell, and S. F. Dice. 1967. Distribution and cycling of nitrogen, phosphorus, potassium, and calcium in a second growth Douglas fir ecosystem. In H. E. Young (ed.), *Symposium on primary productivity and mineral cycling in natural ecosystems*, pp. 197–232. College of Life Sciences and Agriculture, University of Maine, Orono, Maine.

Cole, D. W., W. J. B. Crane, and C. C. Grier. 1975. The effect of forest management practices on water chemistry in a second-growth Douglas-fir ecosystem. In B. Bernier, and C. H. Winget (eds), *Forest soils and forest land management*, Proceedings of the Fourth North American Forest Soils Conference, pp. 195–207. Les Presses de L'Université Laval, Quebec.

Cole, D. W., and D. W. Johnson. Undated. Mineral cycling in tropical forests. Publication number 1269, Environmental Sciences Division, Oak Ridge National Laboratory, Oak Ridge, Tennessee.

Cole, D. W., and M. Rapp, 1981. Elemental cycling in forest ecosystems. In D. E. Reichle (ed.), *Dynamic properties of forest ecosystems*, pp. 341–409. Cambridge University Press, Cambridge.

Coleman, D. C. 1976. A review of root production processes and their influence on soil biota in terrestrial ecosystems. In J. M. Anderson and A. MacFadyen (eds) *The role of terrestrial and aquatic organisms in decomposition processes*. The 17th Symposium of the British Ecological Society. pp. 417–434. Blackwell Scientific, Oxford.

Coleman, D. C., C. P. P. Reid, and C. V. Cole. 1983. Biological strategies of nutrient cycling in soil systems. *Advances in Ecological Research*, **13**: 1–55.

Coleman, N. T., and G. W. Thomas. 1967. The basic chemistry of soil acidity. In R. W. Pearson, and F. Adams (eds), *Soil acidity and liming. (Agronomy Monographs*, **12**: 1–41).

Coley, P. D. 1982. Rates of herbivory on different tropical trees. In E. G. Leigh, A. S. Rand, and D. M. Windsor (eds), *The ecology of a tropical forest*, pp. 123–132. Smithsonian Institution Press. Washington.

Coley, P. D. 1983. Herbivory and defensive characteristics of tree species in a lowland tropical forest. *Ecological Monographs*, **53**: 209–233.

Comar, C. L., and F. W. Lengemann. 1967. General principles of the distribution and movement of artificial fallout through the biosphere to man. In B. Aberg and F. P. Hungate (eds), *Radioecological concentration processes*, Proceedings of an International Symposium, pp. 1–18. Pergamon Press, Oxford.

Connell, J. H., J. G. Tracy, and L. J. Webb. 1984. Compensatory recruitment, growth, and mortality as factors maintaining rain forest tree diversity. *Ecological Monographs*, **54**: 141–164.

Crow, T. R. 1980. A rainforest chronical: a 30 year record of changes in structure and composition at El Verde, Puerto Rico. *Biotropica*, **12**: 42–55.

Dalton, J. D., G. C. Russell, and D. H. Sieling. 1952. Effect of organic matter on phosphate availability. *Soil Science*, **73**: 173–181.

De las Salas, G. (ed.). 1979. *Proceedings, workshop on agro-forestry systems in Latin America, Turrialba, Costa Rica, March 1979*. Centro Agronomico Tropical de Investigacion y Enseñanza, Turrialba, Costa Rica.

Dean, J. M., and A. P. Smith. 1978. Behavioral and morphological adaptations of a tropical plant to high rainfall. *Biotropica*, **10**: 152–154.

Deevey, E. S. 1970. Mineral cycles. *Scientific American*, **223**: 149–158.

Delmas, R., J. Baudet, J. Servant, and Y. Baziard. 1980. Emissions and concentrations of hydrogen sulfide in the air of the tropical forest of the Ivory Coast and temperate regions of France. *Journal of Geophysical Research*, **85**(C8): 4468–4474.

Delwiche, C. C. 1970. The nitrogen cycle. *Scientific American*, **223** (3): 136–146.

Delwiche, C. C. 1977. Energy relations in the global nitrogen cycle. *Ambio*, **6**: 106–111.

Denevan, W. M. 1971. Campa subsistence in the Gran Pajonal, eastern Peru. *The Geographical Review*, **61**: 496–518.

Denevan, W. M. 1981. Swiddens and cattle versus forest: the imminent demise of the Amazon rain forest reexamined. In V. H. Sutlive, N. Altshuler, and M. D. Zamora (eds), *Where have all the flowers gone? Deforestation in the Third World*, pp. 25–44. Publication no. 13, Studies in Third World Societies, Department of Anthropology, College of William and Mary, Williamsburg, Va.

D'hoore, J., and J. K. Coulter. 1972. Soil silicon and plant nutrition. In *Soils of the humid tropics*, pp. 163–173. National Academy of Sciences, Washington, D.C.

160

Diaz del Castillo, B. 1916. *The true history of the conquest of New Spain*, translated by A. P. Maudslay; G. Garcia, (ed.) Hakluyt Society, London.

Döbereiner, J. 1977. Biological nitrogen fixation in tropical grasses – possibilities for partial replacement of mineral N fertilizers. *Ambio*, **6**: 174–177.

Dowler, C. C., and F. H. Tschirley. 1970. Effects of herbicides on a Puerto Rican rain forest. In H. T. Odum and R. F. Pigeon (eds), *A tropical rain forest*, pp. B-315–B-323. Division of Technical Information, US Atomic Energy Commission, Washington, D.C.

Drapcho, D. L., D. Sisterson, and R. Kumar. 1983. Nitrogen fixation by lightning activity in a thunderstorm. *Atmospheric Environment*, **17**: 729–734.

Drew, W. B., S. Aksornkoae, and W. Kaitpraneet. 1978. An assessment of productivity in successional stages from abandoned swidden (rai) to dry evergreen forest in northeastern Thailand. Forest Research Bulletin No. 56, Faculty of Forestry, Kasetsart University, Bangkok, Thailand.

Duvigneaud, P., and S. Denaeyer-DeSmet. 1970. Biological cycling of minerals in temperate deciduous forests. In D. E. Reichle (ed.), *Analysis of temperate forest ecosystems*, pp. 199–225. Springer-Verlag, New York.

Edmisten, J. 1970. Soil studies in the El Verde rain forest. In H. T. Odum and R. F. Pigeon (eds), *A tropical rain forest*, pp. H-79–H-87. Division of Technical Information, US Atomic Energy Commission, Washington, D. C.

Edwards, P. J. 1977. Studies of mineral cycling in a montane rain forest in New Guinea. II. The production and disappearance of litter. *Journal of Ecology*, **65**: 971–992.

Edwards, P. J. 1982. Studies of mineral cycling in a montane rain forest in New Guinea. V. Rates of cycling in throughfall and litter fall. *Journal of Ecology*, **70**: 807–827.

Edwards, P. J., and P. J. Grubb. 1977. Studies of mineral cycling in a montane rain forest in New Guinea. I. The distribution of organic matter in the vegetation and soil. *Journal of Ecology*, **65**: 943–969.

Edwards, P. J., and P. J. Grubb. 1982. Studies of mineral cycling in a montane rain forest in New Guinea. IV. Soil characteristics and the division of mineral elements between the vegetation and soil. *Journal of Ecology*, **70**: 649–666.

Engelberg, J., and L. L. Boyarsky. 1979. The noncybernetic nature of ecosystems. *American Naturalist*, **114**: 317–324.

Evans, T. D., and J. K. Syers. 1971. An application of autoradiography to study the spacial distribution of ^{32}P-labelled orthophosphate added to soil crumbs. *Soil Science Society of America Proceedings*, **35**: 906–909.

Ewel, J. 1971. Biomass changes in early tropical succession. *Turrialba*, **21**: 110–112.

Ewel, J. 1980. Tropical succession: manifold routes to maturity. *Biotropica*, **12** (Supplement): 2–7.

Ewel, J., F. Benedict, C. Berish, and B. Brown. 1982. Leaf area, light transmission, roots and leaf damage in nine tropical plant communities. *Agro-Ecosystems*, **7**: 305–326.

Ewel, J., C. Berish, B. Brown, N. Price, and J. Raich. 1981. Slash and burn impacts on a Costa Rican wet forest site. *Ecology*, **62**: 816–829.

Ewel, K. C., J. F. Gamble, and A. E. Lugo. 1975. Aspects of mineral-nutrient cycling in a southern mixed-hardwood forest in north central Florida. In F. G. Howell, J. B. Gentry, and M. H. Smith (eds), *Mineral cycling in southeastern ecosystems*, pp. 700–714. CONF 740513, Technical Information Center, US Energy Research and Development Administration, Washington, D.C.

Falesi, I. C. 1976. Ecossistema de pastegem cultivada na Amazônia Brasileira. Boletim Técnico No. 1, EMBRAPA Centro de Pesquisa Agropecuaria do Trópico úmido, Belém-Pará, Brasil.

Fearnside, P. M. 1978. Estimation of carrying capacity for human populations in a part of the Transamazon Highway colonization area of Brazil. PhD dissertation, Department of Biology, University of Michigan.

161

Fearnside, P. M., and J. Rankin. 1980. Jarí and development in the Brazilian Amazon. *Interciencia*, **5**: 146–156.

Fearnside, P. M., and J. M. Rankin. 1982. The new Jarí: risks and prospects of a major Amazonian development. *Interciencia*, **7**: 329–339.

Finn, J. T. 1976. Measures of ecosystem structure and function derived from analysis of flows. *Journal of Theoretical Biology*, **56**: 363–380.

Finn, J. T. 1978. Cycling index: a general definition for cycling in compartment models. In D. C. Adriano and I. L. Brisbin (eds), *Environmental chemistry and cycling processes*, pp. 138–164. CONF-760429, Technical Information Center, US Department of Energy, Washington, D.C.

Fölster, H., G. De las Salas, and P. Khanna. 1976. A tropical evergreen forest site with perched water table, Magdalena Valley, Columbia. Biomass and bioelement inventory of primary and secondary vegetation. *Oecologia Plantarum*, **11**: 297–320.

Food and Agriculture Organization. 1974. *Soil map of the world*, UNESCO, Paris.

Forman, R.T. 1975. Canopy lichens with blue-green algae: a nitrogen source in a Colombian rain forest. *Ecology*, **56**: 1176–1184.

Forman, R. T. (ed.). 1979. *Pine barrens: ecosystem and landscape*. Academic Press. New York.

Foster, N. W., and I. K. Morrison. 1976. Distribution and cycling of nutrients in a natural *Pinus banksiana* ecosystem. *Ecology*, **57**: 110–120.

Fox, J. E. D. 1976. Constraints on the natural regeneration of tropical moist forest. *Forest Ecology and Management*, **1**: 37–65.

Fox, R. L., and P. G. E. Searle. 1978. Phosphate adsorption by soils of the tropics. In M. Stelly (ed.), *Diversity of soils in the tropics*, pp. 97–119. ASA Special Publication No. 34, American Society of Agronomy, Madison, Wisc.

Fox, R. L., J. A. Silva, O. R. Younge, D. L. Plucknett, and G. D. Sherman. 1967. Soil and plant silicon and silicate response by sugar cane. *Soil Science Society of America Proceedings*, **31**: 775–779.

Francis, P., and S. Self. 1983. The eruption of Krakatau. *Scientific American*, **249** (5): 172–187.

Frangi, J. L., and A. E. Lugo. 1985. Structure and carbon, phosphorus, and water dynamics of a subtropical floodplain forest. *Ecology*, in press.

Frieden, E. 1972. The chemical elements of life. *Scientific American*, **227** (7): 52–60.

Friedman, I. 1977. The Amazon Basin, another Sahel? *Science*, **197**: 15.

Fuentes Flores, R. 1979. Coffee production farming systems in Mexico. In G. De las Salas (ed.), *Workshop on agro-forestry systems in Latin-America, Turrialba, Costa Rica, March 1979*, pp. 60–72. Centro Agronomico Tropical de Investigacion y Enseñanza, Turrialba, Costa Rica.

Fyfe, W. S., B. I. Kronberg, O. H. Leonardos, and N. Olorunfemi. 1983. Global tectonics and agriculture: a geochemical perspective. *Agriculture, Ecosystems and Environment*, **9**: 383–399.

Galbally, I. E., J. A. Garland, and M. J. G. Wilson. 1979. Sulfur uptake from the atmosphere by forest and farmland. *Nature*, **280**: 49–50.

Galbally, I. E., G. D. Farquhar, and G. P. Ayers. 1982. Interactions in the atmosphere of the biogeochemical cycles of carbon, nitrogen, sulfur. In J. R. Freney, and I. E. Galbally (eds), *Cycling of carbon, nitrogen, sulfur, and phosphorus in terrestrial and aquatic ecosystems*, pp. 1–9. Springer-Verlag, Berlin.

Galston, A. W., and P. W. Richards. 1984. Terrestrial plant ecology and forestry: an overview. In A. H. Westing (ed.), *Herbicides in war, the long-term ecological and human consequences*, pp. 39–42. Taylor and Francis, London.

Gamble, J. F., and S. C. Snedaker. 1969. Final report, Agricultural ecology, IOCS Memo BMI-30, Battelle Memorial Institute, Columbus, Ohio.

162

Gamble, T. N., M. R. Betlach, and J. M. Tiedje. 1977. Numerically dominant denitrifying bacteria from world soils. *Applied and Environmental Microbiology*, **33**: 926–939.

Garwood, N. C., D. P. Janos, and N. Brokow. 1979. Earthquake-caused landslides: a major disturbance to tropical forests. *Science*, **205**: 997–999.

Gauch, H. G. 1977. Ordiflex – a flexible computer program for four ordination techniques. Ecology and Systematics, Cornell University, Ithaca, N.Y.

Gauch, H. G., R. H. Whittaker, and T. R. Wentworth. 1977. A comparative study of reciprocal averaging and other ordination techniques. *Journal of Ecology*, **65**: 157–174.

Gerloff, G. C., and W. H. Gabelman. 1983. Genetic basis of inorganic plant nutrition. In A. Lauchli and R. L. Bielski (eds), *Encyclopedia of plant physiology*, New Series, Vol. 15B, *Inorganic Plant Nutrition*, pp. 453–480. Springer-Verlag, Berlin.

Gessel, S. P., D. W. Cole, D. Johnson, and J. Turner. 1977. The nutrient cycles of two Costa Rican forests. In *Actas del IV Symposium Internacional de Ecologia Tropical, March 1977, Republic of Panama*, Vol. II, pp. 623–643. University of Panama, Panama City.

Gilbert, L. E. 1980. Food web organization and the conservation of neotropical diversity. In M. E. Soule, and B. A. Wilcox (eds), *Conservation biology: an evolutionary—ecological perspective*, pp. 11–33. Sinauer, Sunderland, Mass.

Gile, L. H., F. F. Peterson, and R. B. Grossman. 1966. Morphological and genetic sequences of carbonate accumulation in desert soils. *Soil Science*, **101**: 347–360.

Gliessman, S. R., R. Garcia, and M. Amador. 1981. The ecological basis for the application of traditional agricultural technology in the management of tropical agro-ecosystems. *Agro-Ecosystems*, **7**: 173–185.

Golley, F. B., J. T. McGinnis, R. G. Clements, G. I. Child, and M. J. Deuver. 1975. *Mineral cycling in a tropical moist forest ecosystem*. University of Georgia Press, Athens, Ga.

Gómez-Pompa, A., and C. Vazquez-Yanes. 1974. Studies on the secondary succession of tropical lowlands: the life cycle of secondary species. In *Proceedings of the first international congress of ecology, structure, functioning and management of ecosystems, The Hague, The Netherlands, September 8—14, 1974*, pp. 336–342.

Gómez-Pompa, A., C. Vazquez-Yanes, and S. Guevara. 1972. The tropical rain forest: a nonrenewable resource. *Science*, **177**: 762–765.

Goodland, R. J. A. 1980. Environmental ranking of Amazonian development projects in Brazil. *Environmental Conservation*, **7**: 9–26.

Goodland, R. J. A., and H. S. Irwin. 1975. *Amazon jungle: green hell to red desert?* Elsevier, Amsterdam.

Goodland, R., and R. Pollard. 1973. The Brazilian cerrado vegetation: a fertility gradient. *Journal of Ecology*, **61**: 219–224.

Gordon, B. L. 1982. *A Panama forest and shore*. Boxwood Press, Pacific Grove, Cali.

Gosz, J. R. 1975. Nutrient budgets for undisturbed ecosystems along an elevational gradient in New Mexico. In F. G. Howell, J. B. Gentry, and M. H. Smith (eds), *Mineral cycling in southeastern ecosystems*, pp. 780–799. CONF 740513, Technical Information Center, US Energy Research and Development Administration, Washington, D.C.

Gosz, J. R., and F. M. Fisher. 1984. Microbial responses to ecosystem perturbations. In M. J. Klug and C. A. Reddy (eds), *Current perspectives in microbial ecology*, pp. 523–530. American Society for Microbiology, Washington, D.C.

Graustein, W. C., K. Cromack, and P. Sollins. 1977. Calcium oxalate: occurrence in soils and effect on nutrient and geochemical cycles. *Science*, **198**: 1252–1254.

Greaves, A. 1979. *Gmelina* large scale planting, Jarilandia, Amazon Basin. *Commonwealth Forestry Review*, **58**: 267–269.

Greenland, D. J. 1958. Nitrate fluctuations in tropical soils. *Journal Agricultural Science*, **50**: 82–92.

Greenland, D. J. 1975. Bringing the green revolution to the shifting cultivator. *Science*, **190**: 841–844.

Greenland, D. J., and P. H. Nye. 1959. Increases in the carbon and nitrogen contents of tropical soils under natural fallows. *Soil Science*, **10**: 284–299.

Greenland, D. J., and J. M. L. Kowal. 1960. Nutrient content of the moist tropical forest of Ghana. *Plant and Soil*, **12**: 154–175.

Grimm, U., and H. W. Fassbender. 1981. Ciclos bioquimicos en un ecosistema forestal de los Andes Occidental de Venezuela. *Turrialba*, **31**: 27–37.

Grubb, P. J. 1977. Control of forest growth and distribution on wet tropical mountains, with special reference to mineral nutrition. *Annual Review of Ecology and Systematics*, **8**, 83–107.

Grubb, P. J., and P. J. Edwards. 1982. Studies of mineral cycling in a montane rain forest in New Guinea. III. The distribution of mineral elements in the aboveground material. *Journal of Ecology*, **70**: 623–648.

Hammond, N. 1982. The exploration of the Maya world. *American Scientist*, **70**: 482–495.

Harcombe, P. A. 1977a. The influence of fertilization on some aspects of succession in a humid tropical forest. *Ecology*, **58**: 1375–1383.

Harcombe, P. A. 1977b. Nutrient accumulation by vegetation during the first year of recovery of a tropical forest ecosystem. In J. Cairns, K. L. Dickson, and E. E. Herricks (eds), *Recovery and restoration of damaged ecosystems*, pp. 347–378. University of Virginia Press, Charlottesville.

Harris, D. R. 1971. The ecology of swidden cultivation in the upper Orinoco rain forest, Venezuela. *The Geographical Review*, **61**: 475–495.

Harris, W. F., R. S. Kinerson, and N. T. Edwards. 1977. Comparison of belowground biomass of natural deciduous forests and loblolly pine plantations. In J. K. Marshall (ed.), *The belowground ecosystem: a synthesis of plant-associated processes*, pp. 29–37. Range Science Department Science Series No. 26, Colorado State University, Fort Collins, Colo.

Hart, R. D. 1980. A natural ecosystem analog approach to the design of a successional crop system for tropical forest environments. *Biotropica*, **122** (Supplement): 73–82.

Hartshorn, G. S. 1978. Tree falls and tropical forest dynamics. In P. B. Tomlinson, and M. H. Zimmerman (eds), *Tropical trees as living systems*, pp. 617–638. Cambridge University Press, Cambridge.

Hartshorn, G. S. 1981. Report to Institute of Current World Affairs, December 1981, on activities as a Forest and Man Fellow, sponsored by that Institute. Institute of Current World Affairs, Wheelock House, Hanover, N.H.

Hase, H., and H. Fölster. 1982. Bioelement inventory of a tropical (semi-) evergreen seasonal forest on eutrophic alluvial soils, Western Llanos, Venezuela. *Acta Oecologia, Oecologia Plantarum*, **3**: 331–346.

Hecht, S. B. 1981. Deforestation in the Amazon Basin: magnitude, dynamics and soil resource effects. In V. H. Sutlive, N. Altshuler, and M. D. Zamora (eds), *Where have all the flowers gone? Deforestation in the Third World*, pp. 61–108. Publication No. 13, Studies in Third World Societies, Department of Anthropology, College of William and Mary, Williamsburg, Va.

Heinrichs, H., and R. Mayer. 1977. Distribution and cycling of major and trace elements in two central European forest ecosystems. *Journal of Environmental Quality*, **6**: 402–406.

Henderson, G. S., N. T. Edwards, D. E. Reichle, C. W. Francis, M. H. Shanks, and P. Sollins. 1971. Mineral cycling. In *Ecological Sciences Division, annual progress report*, pp. 85–90. Oak Ridge National Laboratory, Oak Ridge, Tennessee.

Henderson, G. S., A. Hunley, and W. Selvidge. 1977. Nutrient discharge from Walker Branch Watershed. In D. L. Correll (ed.), *Watershed research in eastern North America*, pp. 307–321. Chesapeake Bay Center for Environmental Studies, Edgewater, Md.

Hermann, R. K. 1977. Growth and production of tree roots: a review. In J. K. Marshall (ed.), *The belowground ecosystem: a synthesis of plant-associated processes*, pp. 7–28. Range Science Department Science Series No. 26, Colorado State University, Fort Collins, Colo.

Herrera, R. A. 1979. Nutrient distribution and cycling in an Amazon caatinga forest on Spodosols in southern Venezuela. PhD dissertation, University of Reading, England.

Herrera, R., T. Merida, N. Stark, and C. Jordan. 1978a. Direct phosphorus transfer from leaf litter to roots. *Naturwissenschaften*, **65**: 208–209.

Herrera, R., C. F. Jordan, H. Klinge, and E. Medina. 1978b. Amazon ecosystems: their structure and functioning with particular emphasis on nutrients. *Interciencia*, **3**: 223–232.

Heslop-Harrison, Y. 1978. Carnivorous plants. *Scientific American*, **238** (2): 104–115.

Hirose, T., and M. Tateno. 1984. Soil nitrogen patterns induced by colonization of *Polygonum cuspidatum* on Mt. Fuji. *Oecologia*, **61**: 218–223.

Hodges, C. S. 1981. The implications of diseases and insects in managing tropical forests. In M. Chavarria (ed.), *Simposio Internacional sobre las ciencias forestales y su contribucion al desarrollo de la America tropical, San Jose, Costa Rica, Octubre 1979*, pp. 73–80. CATIE, Turrialba, Costa Rica.

Holdridge, L. R. 1967. *Life zone ecology*. Tropical Science Center, San Jose, Costa Rica.

Hornbeck, J. W., G. E. Likens, R. S. Pierce, and F. H. Bormann. 1975. Strip cutting as a means of protecting site and streamflow quality when clearcutting northern hardwoods. In B. Bernier and C. H. Winget (eds), *Forest soils and forest land management*. Proceedings of the Fourth North American Forest Soils Conference, pp. 209–225. Les Presses de l'Université, Laval, Quebec.

Howeler, R. H., L. F. Cadavid, and E. Burckhardt. 1982. Response of cassava to VA mycorrhizal inoculation and phosphorus application in greenhouse and field experiments. *Plant and Soil*, **69**: 327–339.

Hubbell, S. P. 1979. Tree dispersion, abundance, and diversity in a tropical dry forest. *Science*, **203**: 1299–1309.

Ivanov, M. V. 1981. The global biogeochemical sulphur cycle. In G. E. Likens (ed.), *Some Perspectives of the Major Biogeochemical Cycles*, pp. 61–78. Wiley, Chichester.

Jackson, M. L. 1964. Chemical composition of soils. In F. E. Bear (ed.), *Chemistry of the Soil*, pp. 71–141. Reinhold, New York.

Jackson, M. L., S. A. Tyler, A. L. Willis, G. A. Bourdeau, and R. P. Pennington. 1948. Weathering sequence of clay-sized minerals in soils and sediments. *Journal of Physical and Colloidal Chemistry*, **52**: 1237–1260.

Jackson, R. M., and F. Raw. 1973. *Life in the soil*. Edward Arnold, London.

Jaiyebo, E. O., and A. W. Moore. 1964. Soil fertility and nutrient storage in different soil–vegetation systems in a tropical rainforest environment. *Tropical Agriculture, Trinidad*, **41**: 129–139.

Janos, D. P. 1980a. Mycorrhizae influence tropical succession. *Biotropica*, Supplement to Vol. **12** (2): 56–64.

Janos, D. P. 1980b. Vesicular-arbuscular mycorrhizae affect lowland tropical rain forest plant growth. *Ecology*, **61**: 151–162.

Janos, D. P. 1983. Tropical mycorrhizas, nutrient cycles, and plant growth. In S. L. Sutton, T. C. Whitmore, and A. C. Chadwick (eds), *Tropical rain forest: ecology and management*, pp. 327–345. Blackwell Scientific, Oxford.

Janson, C. H. 1983. Adaptation of fruit morphology to dispersal agents in a neotropical forest. *Science*, **219**: 187–189.

Janzen, D. H. 1973. Tropical agroecosystems. *Science*, **182**: 1212–1219.

Janzen, D. H. 1974. Tropical blackwater rivers, animals, and mast fruiting by the Dipterocarpaceae. *Biotropica*, **6**: 69–103.

Janzen, D. H. 1983. Food webs: who eats what, why, how and with what effects in a tropical forest? In F. B. Golley (ed.), *Ecosystems of the world*, Vol. 14A, *Tropical rain forest ecosystems: structure and function*, pp. 167–182. Elsevier, Amsterdam.

Jenny, H. 1950. Causes of the high nitrogen and organic matter content of certain tropical forest soils. *Soil Science*, **69**: 63–69.

Jenny, H. 1980. *The soil resource* (Ecological Studies, Vol. 37), Springer-Verlag, New York.

Jenny, H., S. P. Gessel, and F. T. Bingham. 1949. Comparative study of decomposition rates of organic matter in temperate and tropical regions. *Soil Science*, **68**: 419–432.

Johnson, D. W., D. W. Cole, and S. P. Gessel. 1975. Processes of nutrient transfer in a tropical rain forest. *Biotropica*, **7**: 208–215.

Johnson, D. W., D. W. Cole, S. P. Gessel, M. J. Singer, and R. V. Minden. 1977. Carbonic acid leaching in a tropical, temperate, subalpine, and northern forest soil. *Arctic and Alpine Research*, **9**: 329–343.

Johnson, D. W., D. W. Cole, and S. P. Gessel. 1979. Acid precipitation and soil sulfate adsorption properties in a tropical and in a temperate forest soil. *Biotropica*, **11**: 38–42.

Johnson, D. W., G. S. Henderson, D. D. Huff, S. E. Lindberg, D. D. Richter, D. S. Shriner, D. E. Todd, and J. Turner. 1982a. Cycling of organic and inorganic sulphur in a chestnut oak forest. *Oecologia*, **54**: 141–148.

Johnson, D. W., D. W. Cole, C. S. Bledsoe, K. Cromack, R. L. Edmonds, S. P. Gessel, C. C. Grier, B. N. Richards, and K. A. Vogt. 1982b. Nutrient cycling in forests of the Pacific Northwest. In R. L. Edmonds (ed.), *Analysis of coniferous forest ecosystems in the western United States*, pp. 186–232. US/IBP Synthesis Series 14, Hutchinson Ross, Stroudsburg, Penn.

Johnson, D. W., D. C. West, D. E. Todd, and L. K. Mann. 1982c. Effects of sawlog vs. whole-tree harvesting on the nitrogen, phosphorus, potassium, and calcium budgets of an upland mixed oak forest. *Soil Science Society of America Journal*, **46**: 1304–1309.

Johnson, F. L., and P. G. Risser, 1974. Biomass, annual net primary production, and dynamics of six mineral elements in a post oak-blackjack oak forest. *Ecology*, **55**: 1246–1258.

Johnson, P. L., and W. T. Swank. 1973. Studies of cation budgets in the Southern Appalachians on four experiment watersheds with contrasting vegetation. *Ecology*, **54**: 70–80.

Jordan, C. F. 1969. Derivation of leaf-area index from quality of light on the forest floor. *Ecology*, **50**: 663–666.

Jordan, C. F. 1971. Productivity of a tropical forest and its relation to a world pattern of energy storage. *Journal of Ecology*, **59**: 127–142.

Jordan, C. F. 1982a. The nutrient balance of an Amazonian rain forest. *Ecology*, **63**: 647–654.

Jordan, C. F. 1982b. Nutrient cycling index of an Amazonian rain forest. *Acta Oecologia, Oecologia Generalis*, **3**: 393–400.

Jordan, C. F. 1982c. Amazon rain forests. *American Scientist*, **70**: 394–401.

Jordan, C. F. 1983. Productivity of tropical rain forest ecosystems and the implications for their use as future wood and energy sources. In F. B. Golley (ed.), *Ecosystems of the world*, Vol. 14A, *Tropical rain forest ecosystems: structure and function*, pp. 117–135. Elsevier, Amsterdam.

Jordan, C. F. 1985. An Amazonian ecosystem. Manuscript submitted.

Jordan, C. F., and E. G. Farnworth. 1980. A rain forest chronicle: perpetuation of a myth. *Biotropica*, **12**: 233–234.

Jordan, C. F., and E. G. Farnworth. 1982. Natural vs. plantation forests: a case study of land reclamation strategies for the humid tropics. *Environmental Management*, **6**: 485–492.

Jordan, C. F., and R. Herrera. 1981. Tropical rain forests: are nutrients really critical? *American Naturalist*, **117**; 167–180.

Jordan, C. F., and J. R. Kline. 1976. Strontium-90 in a tropical rain forest: 12th-year validation of a 32-year prediction. *Health Physics*, **30**: 199–201.

Jordan, C. F., and P. G. Murphy. 1978. A latitudinal gradient of wood and litter production and its implication regarding competition and species diversity in trees. *American Midland Naturalist*, **99**: 414–434.

Jordan, C. F., and C. E. Russell. 1983. Jarí: productividad de las plantaciones y perdida de nutrientes debido al corte yla quema. *Interciencia*, **8**: 294–297.

Jordan, C. F. and C. Uhl. 1978. Biomass of a 'tierra firme' forest of the Amazon Basin. *Oecologia Plantarum*, **13**: 387–400.

Jordan, C. F., J. R. Kline, and D. S. Sasscer. 1972. Relative stability of mineral cycles in forest ecosystems. *American Naturalist*, **106**: 237–253.

Jordan, C. F., J. R. Kline, and D. S. Sasscer. 1973. A simple model of strontium and manganese dynamics in a tropical rain forest. *Health Physics*, **24**: 477–489.

Jordan, C. F., R. L. Todd, and G. S. Escalante. 1979a. Nitrogen conservation in a tropical rain forest. *Oecologia*, **39**: 123–128.

Jordan, C. F., F. Golley, J. D. Hall, and J. Hall. 1979b. Nutrient scavenging of rainfall by the canopy of an Amazonian rain forest. *Biotropica*, **12**: 61–66.

Jordan, C., W. Caskey, G. Escalante, R. Herrera, F. Montagnini, R. Todd, and C. Uhl. 1982. The nitrogen cycle in a 'tierra firme' rainforest on oxisol in the Amazon Territory of Venezuela. *Plant and Soil*, **67**: 325–332.

Jordan, E. L. 1981. The birth of ecology. In C. F. Jordan (ed.), *Benchmark papers in tropical ecology*, pp. 4–15. Hutchinson Ross, Stroudsburg, Penn.

Kang, B. T., and R. Lal. 1981. Nutrient losses in water runoff from agricultural catchments. In R. Lal and E. W. Russell (eds), *Tropical agricultural hydrology: watershed management and land use*, pp. 153–161. Wiley, New York.

Karel, A. K., D. A. Lakhani, and B. J. Ndunguru. 1982. Intercropping of maize and cowpea: effect of plant populations on insect pests and seed yield. In C. L. Keswani, and B. J. Ndunguru (eds), *Intercropping*, Proceedings of the Second Symposium on Intercropping, Morogoro, Tanzania, pp. 102–109. International Development Research Centre, Ottawa.

Kato, R., Y. Tadaki, and H. Ogawa. 1978. Plant biomass and growth increment studies in Pasoh forest. *Malaysian Nature Journal*, **30**: 211–224.

Kaye, S. V., and S. J. Ball. 1969. Systems analysis of a coupled compartment model for radionuclide transfer in a tropical environment. In D. J. Nelson and F. C. Evans (eds), *Symposium on Radioecology*, Proceedings of the Second National Symposium, pp. 731–739. CONF. 670–503. US Department of Commerce, Springfield, Va.

Kellog, C. E. 1963. Shifting cultivation, *Soil Science*. **95**: 221–230.

Kellog, W. W., R. D. Cadle, E.R. Allen, A. L. Lazrus, and E. A. Martell. 1972. The sulfur cycle. *Science*, **175**: 587–596.

Kenworthy, J. B. 1971. Water and nutrient cycling in a tropical rain forest. In J. R. Flenly (ed.), *The water relations of Malesian forests*, Transactions of the First Aberdeen–Hull Symposium on Malesian Ecology, pp. 49–59. Institute for Southeast Asian Biology, University of Aberdeen, Aberdeen.

Keswani, C. L., and B. J. Ndunguru (eds). 1982. *Intercropping*, Proceedings of the Second Symposium on Intercropping, Morogoro, Tanzania. International Development Research Centre. Ottawa.

Kimmins, J. P., and G. J. Krumlik. 1976. On the question of nutrient losses accompanying whole tree logging. In *Oslo biomass studies*, pp. 43–53. International Union of Forest Research Organizations, University of Maine, Orono, Maine.

Kinkead, G. 1981. Trouble in D. K. Ludwig's jungle. *Fortune*, April 20, 1981: 102–117.

Klinge, H. 1976. Bilanzierung von Hauptnahrstoffen im Ökosystem tropischer regenwald (Manaus) – vorlaufige daten. *Biogeographica*, **7**: 59–76.

Klinge, H., and E. J. Fittkau. 1972. Filterfunktionen im Okosystem des zentralamazonischen regenwaldes. *Mitteilungen der deutschen bodenkundliche Gesellschaft*, **16**: 130–135.

Klinge, H., and R. Herrera. 1978. Biomass studies in an Amazon caatinga forest in southern Venezuela. I. Standing crop of composite root mass in selected stands. *Tropical Ecology*, **19**: 93–110.

Klinge, H., and E. Medina. 1979. Rio Negro caatingas and campinas, Amazonas states of Venezuela and Brazil. In R. L. Specht (ed.), *Ecosystems of the world*, Vol. 9A, *Heathlands and related shrublands*, pp. 483–488. Elsevier, Amsterdam.

Koeppen, R. C. 1978. Some anatomical characteristics of tropical woods. In *Proceedings of conference on improved utilization of tropical forests*, pp. 69–82. Forest Products Laboratory, Forest Service, US Department of Agriculture.

Koranda, J. J. 1965. Preliminary studies of the persistence of tritium and ^{14}C in the Pacific proving ground. *Health Physics*, **11**: 1445–1457.

Kronberg, B. I., W. S. Fyfe, O. H. Leonardos, and A. M. Santos. 1979. The chemistry of some Brazilian soils: element mobility during intense weathering. *Chemical Geology*, **24**: 211–229.

Kronberg, B. I., W. S. Fyfe, B. J. McKinnon, J. F. Couston, B. Stilianidi-Filho, and R. A. Nash. 1982. Model for bauxite formation: Paragominas (Brazil). *Chemical Geology*, **35**: 311–320.

Kuo Keel, S. H., and G. T. Prance. 1979. Studies of the vegetation of a white-sand black-water igapo (Rio Negro, Brazil). *Acta Amazonica*, **9**: 645–655.

Lal, R. 1981a. Clearing of a tropical forest. II. Effects on crop performance. *Field Crops Research*, **4**: 345–354.

Lal, R. 1981b. Deforestation of tropical rainforest and hydrological problems. In R. Lal and E. W. Russell (eds), *Tropical agricultural hydrology: watershed management and land use*, pp. 131–140. Wiley, New York.

Lamb, A. F. A. 1968. Artificial regeneration within the humid lowland tropical forest. *Unasylva*, **22**(4): 7–15.

Lamb, D. 1980. Soil nitrogen mineralization in a secondary rainforest succession. *Oecologia*, **47**: 257–263.

Lang, G. E., and D. H. Knight. 1979. Decay rate of boles of tropical trees in Panama. *Biotropical*, **11**: 316–317.

Lanly, J. P. 1982. Tropical forest resources. FAO Forestry Paper No. 30, Food and Agriculture Organization of the United Nations, Rome.

Lavelle, P. 1983. The structure of earthworm communities. In J. E. Satchell (ed.), *Earthworm ecology: from Darwin to vermiculture*, pp. 449–466. Chapman and Hall, London.

Leech, R. H., and Y. T. Kim. 1981. Foliar analysis and DRIS as a guide to fertilizer amendments in popular plantations. *Forestry Chronicle*, February, 1981: 17–21.

Lemee, G. Undated. Recherches sur l'ecosysteme de la foret sub-equatoriale de basse Cote-d'Ivoire. Unpublished.

Levin, D. A. 1976. The chemical defenses of plants to pathogens and herbivores. *Annual Review of Ecology and Systematics*, **7**: 121–159.

Lewis, L. A. 1974. Slow movement of earth under tropical rain forest conditions. *Geology*, **2**: 9–10.

Lewis, W. M. 1981. Precipitation chemistry and nutrient loading by precipitation in a tropical watershed. *Water Resources Research*, **17**: 169–181.

Liden, K., and M. Gustafsson. 1967. Relationships and seasonal variation of ^{137}Cs in lichen, reindeer, and man in northern Sweden 1961–1965. In B. Aberg and F. P. Hungate, (eds), *Radioecological concentration processes: proceedings of an international symposium*, pp. 193–208. Pergamon Press, Oxford.

Lieth, H., and R. H. Whittaker, (eds) 1975. *Primary productivity of the biosphere*. Springer-Verlag, New York.

Likens, G. E., F. H. Bormann, and N. M. Johnson. 1969. Nitrification: importance to nutrient losses from a cutover forested ecosystem. *Science*, **163**: 1205–1206.

Likens, G. E., F. H. Bormann, N. M. Johnson, D. W. Fisher, and R. S. Pierce. 1970. Effects of forest cutting and herbicide treatment on nutrient budgets in the Hubbard Brook watershed-ecosystem. *Ecological Monographs*, **40**: 23–47.

Likens, G. E., F. H. Bormann, R. S. Pierce, J. S. Eaton, and N. M. Johnson. 1977. *Biogeochemistry of a forested ecosystem*. Springer-Verlag, New York.

Lim, M. T. 1978. Litterfall and mineral nutrient content of litter in Pasoh Forest Reserve. *Malaysian Nature Journal*, **30**: 375–380.

Lindell, B., and A. Magi. 1967. Observed levels of ^{137}Cs in Swedish reindeer meat. In B. Aberg and F. P. Hungate (eds), *Radioecological concentration processes: proceedings of an international symposium*, pp. 217–219. Pergamon Press, Oxford.

Lindsay, W. L., and E. C. Moreno. 1960. Phosphate equilibria in soils. *Soil Science Society of America Proceedings*, **24**: 177–182.

Little, E. L. 1970. Relationships of trees of the Luquillo experimental forest. In H. T. Odum and R. F. Pigeon (eds), *A tropical rain forest*, pp. B-47–B-61. Division of Technical Information, U.S. Atomic Energy Commission, Washington, D.C.

Lopez-Hernandez, D., and C. P. Burnham. 1974. The covariance of phosphate sorption with other soil properties in some British and tropical soils. *Journal of Soil Science*, **25**: 196–206.

Loveless, A. R. 1961. A nutritional interpretation of sclerophylly based on differences in the chemical composition of sclerophyllous and mesophytic leaves. *Annals of Botany*, **25**: 168–184.

Loveless, A. R. 1962. Further evidence to support a nutritional interpretation of sclerophylly. *Annals of Botany*, **26**: 551–561.

Lowe, R. G. 1977. Experience with the tropical shelterwood system of regeneration in natural forest in Nigeria. *Forest Ecology and Management*, **1**: 193–212.

Lowman, M. D., and J. D. Box. 1983. Variation in leaf toughness and phenolic content among five species of Australian rain forest trees. *Australian Journal of Ecology*, **8**: 17–25.

Lowrance, R., R. Todd, J. Fail, O. Hendrickson, R. Leonard, and L. Asmussen. 1984. Riparian forests as nutrient filters in agricultural watersheds. *BioScience*, **34**: 374–377.

Ludlow, M. M. 1969. Dry matter production, leaf angle, silica content, and digestibility of tropical pasture grasses. *Journal of the Australian Institute of Agricultural Science*, September 1969: 200–201.

Lugo, A. E., and S. Brown. 1982. Conversion of tropical moist forests: a critique. *Interciencia*, **7**: 89–93.

Lugo, A. E., and P. G. Murphy. 1984. Nutrient standing crop and turnover in a subtropical dry forest in Puerto Rico. Manuscript in preparation.

Lugo, A., M. Brinson, M. Cerame Vivas, C. Gist *et al.* 1974. Tropical ecosystem structure and function. In E. G. Farnworth, and F. B. Golley (eds), *Fragile ecosystems*, pp. 67–111. Springer-Verlag, New York.

Lugo, A. E., J. A. Gonzales-Liboy, B. Cintron, and K. Dugger. 1978. Structure, productivity, and transpiration of subtropical dry forest in Puerto Rico. *Biotropica*, **10**: 278–291.

Lugo, A. E., M. Applefield, D. J. Pool, and R. B. McDonald. 1983. The impact of hurricane David on the forests of Dominica. *Canadian Journal of Forest Research*, **13**: 201–211.

Luse, R. A. 1970. The phosphorus cycle in a tropical rain forest. In H. T. Odum and R. F. Pigeon (eds), *A Tropical Rain Forest*, pp. H-161–H-166. Division of Technical Information, US Atomic Energy Commission, Washington, D.C.

Lüttge, U. 1983. Ecophysiology of carnivorous plants. In O. L. Lange, P. S. Nobel, C. B. Osmond, and H. Ziegler (eds), *Physiological plant ecology*, Vol. III, *Responses to the chemical and biological environment*, pp. 489–517. Springer-Verlag, Berlin.

Luvall, J. C. 1984. Tropical deforestation and recovery: the effects on the evapotranspiration process. PhD dissertation, University of Georgia, Athens, Ga.

Lyr, H., and G. Hoffman. 1967. Growth rates and growth periodicity of tree roots. In J. A. Romberger and P. Mikola (eds), *International Review of Forestry Research*, Vol. 2, pp. 181–235. Academic Press, New York.

MacDonald, L. H. (ed.). 1981. *Agro-forestry in the African Humid Tropics*, Proceedings of a workshop held in Ibadan, Nigeria, 27 April–1 May 1981. The United Nations University, Tokyo, Japan.

Mabberley, D. J. 1983. *Tropical rain forest ecology*. Blackie, Glasgow.

McFee, W. W., and E. L. Stone. 1965. Quantity, distribution, and variability of organic matter and nutrients in a forest podzol in New York. *Soil Science Society of America Proceedings*, **29**: 432–436.

McKey, D., P. G. Waterman, J. S. Gartlan, and T. T. Struhsaker. 1978. Phenolic content of vegetation in two African rain forests: ecological implications. *Science*, **202**: 61–64.

McNaughton, S. J. 1976. Serengeti migratory wildebeest: facilitation of energy flow by grazing. *Science*, **191**: 92–94.

McNeil, M. 1964. Laterite soils. *Scientific American*, **211**: 68–73.

Madge, D. S. 1965. Leaf fall and litter disappearance in a tropical forest. *Pedobiologia*, **5**: 273–288.

Mague, T. H. 1977. Ecological aspects of dinitrogen fixation by blue-green algae. In R. W. F. Hardy and A. H. Gibson, (eds), *A treatise on dinitrogen fixation*, pp. 85–140. Wiley, New York.

Manner, H. I. 1976. The effects of shifting cultivation and fire on vegetation and soils in the montane tropics of New Guinea. PhD dissertation, University of Hawaii, Honolulu.

Manner, H. I., R. R. Thaman, and D. C. Hassall. 1984. Phosphate mining induced vegetation changes on Nauru island. *Ecology*, **65**: 1454–1465.

Manokaran, N. 1980. The nutrient contents of precipitation, throughfall, and stemflow in a lowland tropical rain forest in peninsular Malaysia. *The Malaysian Forester*, **43**: 266–289.

Manshard, W. 1974. *Tropical agriculture*. Longman, London.

Margaris, N. S. 1981. Adaptive strategies in plants dominating Mediterranean-type ecosystems. In F. diCastri, D. W. Goodall, and R. L. Specht (eds), *Ecosystems of the World, Vol. XI, Mediterranean-type shrublands*, pp. 309–315. Elsevier, Amsterdam.

Martin, W. E. 1969. Radioecology and the feasibility of nuclear canal excavation. In D. J. Nelson and F. C. Evans, eds. *Symposium on radioecology*, Proceedings of the Second National Symposium, pp. 9–22. CONF. 670503. US Department of Commerce, Springfield, Va.

Matheny, R. T., and D. L. Gurr. 1979. Ancient hydraulic techniques in the Chiapas highlands. *American Scientist*, **67**: 441–449.

Matson, P. A., and R. D. Boone. 1984. Natural disturbance and nitrogen mineralization: wave-form dieback of mountain hemlock in the Oregon Cascades. *Ecology*, **65**: 1511–1516.

Matson, P. A., and R. H. Waring. 1984. Effects of nutrient and light limitation on mountain hemlock: susceptibility to laminated root rot. *Ecology*, **65**: 1517–1524.

Matsumoto, T. 1976. The role of termites in an equatorial rain forest ecosystem of west Malaysia. *Oecologia*, **22**: 153–178.

Matsumoto, T. 1978. The role of termites in the decomposition of leaf litter in the forest floor of Pasoh study area. *Malaysian Nature Journal*, **30**: 405–413.

Mattson, W. J., and N. D. Addy. 1975. Phytophagous insects as regulators of forest primary production. *Science*, **190**: 515–522.

Maury-Lechon, G. 1982. Régénération forestière sur 25 ha de coupe papetière en forêt dense humide de Guyane francaise. *Comptes rendus hebdomadaire des séances de l'Académie des sciences*, **294**: 975–978.

May, K. W., and R. Misangu. 1980. Some observations on the effects of plant arrangements for intercropping. In C. L. Keswani and B. J. Ndunguru (eds), *Intercropping*, Proceedings of the Second Symposium on Intercropping, Morogoro, Tanzania, pp. 37–42. International Development Research Centre, Ottawa.

Medina, E., and H. Klinge. 1983. Productivity of tropical forests and tropical woodlands. In O. L. Lange, P. S. Nobel, C. B. Osmond, M H. Ziegler (eds), *Physiological plant ecology*, Vol. IV, *Ecosystem processes: mineral cycling, productivity and man's influence*, pp. 281–303. Springer-Verlag, Berlin.

Medina, E., and M. Zelwer. 1972. Soil respiration in tropical plant communities. In P. M. Golley and F. B. Golley (eds), *Tropical ecology with an emphasis on organic productivity*, pp. 245–267. Institute of Ecology, University of Georgia, Athens, Ga.

Medina, E., M. Sobrado, and R. Herrera. 1978. Significance of leaf orientation for leaf temperature in an Amazonian sclerophyll vegetation. *Radiation and Environmental Biophysics*, **15**: 131–140.

Medina, E., H. Klinge, C. Jordan, and R. Herrera. 1980. Soil respiration in Amazonian rain forests in the Rio Negro Basin. *Flora*, **170**: 240–250.

Medina, E., V. Garcia, E. Cuevas, and M. Sobrado. 1984. Functional relationships between leaf structure and mineral nutrient content. In E. Medina, R. Herrera, C. F. Jordan, H. Klinge, and C. Uhl (eds), *Structure and function of Amazonian forest ecosystems in the upper Rio Negro region*. Junk, in preparation.

Meentemeyer, V. Undated. AET values for selected ecological research sites. Geography Department, University of Georgia, Athens, Ga.

Melfi, A. J., C. C. Cerri, B. I. Kronberg, W. S. Fyfe, and B. McKinnon. 1983. Granitic weathering: a Brazilian study. *Journal of Soil Science*, **34**: 841–851.

Meyer, F. H., and D. Gottsche. 1971. Distribution of root tips and tender roots of beech. In H. Ellenberg (ed.), *Integrated experimental ecology: methods and results of ecosystem research in the German Solling project* (Ecological Studies, Vol. 2), pp. 48–52. Springer-Verlag, Heidelberg.

Miettinen, J. K., and E. Hasanen. 1967. [137]Cs in Finnish Lapps and other Finns in 1962–6. In B. Aberg and F. P. Hungate (eds), *Radioecological concentration processes: proceedings of an international symposium*, pp. 221–231. Pergamon Press, Oxford.

Miller, W. J., and M. W. Neathery. 1977. Newly recognized trace mineral elements and their role in animal nutrition. *BioScience*, **27**: 674–679.

Millot, G. 1979. Clay. *Scientific American*, **240**: 109–118.

Mongi, H. O., and P. A. Huxley (eds). 1979. *Soils research in agroforestry*, proceedings of an expert consultation held at Nairobi, March 1979. International Council for Research in Agroforestry, Nairobi, Kenya.

Monk, C. D. 1966. An ecological significance of evergreenness. *Ecology*, **47**: 504–505.

Mooney, H. A., C. Field, and C. Vasquez-Yanes. 1984. Photosynthetic characteristics of wet tropical forest plants. In E. Medina, H. A. Mooney and C. Vazquez-Yanes (eds), *Physiological Ecology of plants of the wet tropics*, pp. 113–128. Junk, The Hague.

Moran, E. F. 1981. *Developing the Amazon*. Indiana University Press. Bloomington, Ind.

Mosse, B., D. P. Stribley, and F. LeTacon. 1981. Ecology of mycorrhizae and mycorrhizal fungi. *Advances in Microbial Ecology*, **5**: 137–210.

Myers, N. 1979. *The sinking ark*. Pergamon Press, Oxford.

Myers, N. 1980. *Conversion of tropical moist forests*. National Academy of Sciences, Washington, D.C.

Nadkarni, N. 1981. Canopy roots: convergent evolution in rainforest nutrient cycles. *Science*, **214**: 1023–1024.

National Research Council. 1982. *Ecological aspects of development in the humid tropics*. National Academy Press, Washington, D.C.

Nevstrueva, M. A., P. V. Ramzaev, A. A. Moiseer, M. S. Ibatullin, and L. A. Teplykh. 1967. The nature of ^{137}Cs and ^{90}Sr transport over the lichen–reindeer–man food chain. In B. Aberg and F. P. Hungate (eds), *Radioecological concentration processes: proceedings of an international symposium*, pp. 209–215. Pergamon Press, Oxford.

Newbold, J. D., R. V. O'Neill, J. W. Elwood, and W. VanWinkle. 1982. Nutrient spiralling in streams: implications for nutrient limitation and invertebrate activity. *American Naturalist*, **120**: 628–652.

Nilsson, S. I., H. G. Miller, and J. D. Miller. 1982. Forest growth as a possible cause of soil and water acidification: an examination of the concepts. *Oikos*, **39**: 40–49.

Norman, C. 1983. Vietnam's herbicide legacy. *Science*, **219**: 1196–1197.

Norman, M. J. T. 1979. *Annual cropping systems in the tropics*. University of Florida Press, Gainesville, Fla.

Nortcliff, S., and J. B. Thornes. 1978. Water and cation movement in a tropical rainforest environment. *Acta Amazonica*, **8**: 245–258.

Novikoff, G. 1983. Desertification by overgrazing. *Ambio*, **12**: 102–105.

Nye, P. H., and D. J. Greenland. 1960. The soil under shifting cultivation. Technical Communication 51, Commonwealth Bureau of Soils, Commonwealth Agricultural Bureaux, Farnham Royal, England.

Nye, P. H., and D. J. Greenland. 1964. Changes in the soil after clearing tropical forest. *Plant and Soil*, **21**: 101–112.

Nye, P. H., and P. B. Tinker. 1977. *Solute movement in the soil—root system*. University of California Press, Berkeley, Calif.

Odum, E. P. 1969. The strategy of ecosystem development. *Science*, **164**: 262–270.

Odum, E. P., and L. J. Biever. 1984. Resource quality, mutualism, and energy partitioning in food chains. *American Naturalist*, **124**: 360–376.

Odum, H. T. 1970a. Summary: an emerging view of the ecological system at El Verde. In H. T Odum and R. F. Pigeon (eds), *A tropical rain forest*, pp. I-191–I-289. Division of Technical Information, US Atomic Energy Commission, Washington, D.C.

Odum, H. T. 1970b. Rain forest structure and mineral cycling homeostasis. In H. T. Odum and R. F. Pigeon (eds), *A tropical rain forest*, pp. H-3–H-52. Division of Technical Information, US Atomic Energy Commission, Washington, D.C.

Odum, H. T., and G. Drewry. 1970. The Cesium source at El Verde. In H. T. Odum and R. F. Pigeon (eds.), *A tropical rain forest*, pp. C-23–C-36. Division of Technical Information, US Atomic Energy Commission, Washington, D.C.

Ogawa, H. 1978. Litter production and carbon cycling in Pasoh forest. *Malaysian Nature Journal*, **30**: 367–373.

Olsen, S. R., and L. A. Dean. 1965. Phosphorus. In C. A. Black (ed.), *Methods of soil analysis*, pp. 1035–1049. American Society of Agronomy, Madison, Wisc.

Olson, J. S. 1963. Energy storage and the balance of producers and decomposers in ecological systems. *Ecology*, **44**: 322–332.

Olson, J. S. 1981. Carbon balance in relation to fire regimes. In *Fire regimes and ecosystem properties*, Proceedings of a conference, Honolulu, Hawaii, December 1978, pp. 327–378. US Department of Agriculture Technical Report WO-26, Washington, D.C.

Olson, R. A., and O. P. Engelstad. 1972. Soil phosphorus and sulfur. In *Soils of the humid tropics*, pp. 82–101. National Academy of Sciences, Washington, D.C.

Olson, R. A., R. B. Clark, and J. H. Bennett. 1981. The enhancement of soil fertility by plant roots. *American Scientist*, **69**: 378–384.

Orians, G., J. Apple, R. Billings, L. Fournier et al. 1974. Tropical population ecology. In E. Farnworth and F. Golley (eds) *Fragile ecosystems*, pp. 5–65. Springer-Verlag, New York.

Parker, G. G. 1983. Throughfall and stemflow in the forest nutrient cycle. *Advances in Ecological Research*, **13**: 57–133.

Parker, G. G. 1985. Nutrient loss and recapture following deforestation of tropical hillslope forests. PhD dissertation, University of Georgia, Athens, Ga.

Pastor, J., and J. G. Bockheim. 1984. Distribution and cycling of nutrients in an aspen–mixed-hardwood–spodosol ecosystem in northern Wisconsin. *Ecology*, **65**: 339–353.

Patten, B. C., and E. P. Odum. 1981. The cybernetic nature of ecosystems. *American Naturalist*, **118**: 886–895.

Paul, E. A., and R. M. N. Kucey. 1981. Carbon flow in plant microbial associations. *Science*, **213**: 473–474.

Peace, W. J. H., and F. D. MacDonald. 1981. An investigation of the leaf anatomy, foliar mineral levels, and water relations of trees of a Sarawak forest. *Biotropica*, **13**: 100–109.

Peck, A. J. 1977. Development and reclamation of secondary salinity. In J. S. Russell and E. L. Greacen (eds), *Soil factors in crop production in a semi-arid environment*, pp. 301–319. University of Queensland Press, St. Lucia, Australia.

Persson, T., E. Baath, M. Clarholm, H. Lundkvist, B. E. Soderstrom, and B. Sohlenius. 1980. Trophic structure, biomass dynamics, and carbon metabolism of soil organisms in a scots pine forest. In T. Persson (ed.), *Structure and function of northern coniferous Forests – An Ecosystem Study. Ecological Bulletins*, (Stockholm) **32**: 419–459.)

Peterjohn, W. T., and D. L. Correll. 1984. Nutrient dynamics in an agricultural watershed: observations on the role of a riparian forest. *Ecology*, **65**: 1466–1475.

Poore, D. 1976. The values of tropical moist forest ecosystems. *Unasylva*, **28** (Nos 112–113): 127–143.

Posey, C. E. 1980. Statement of Dr Clayton E. Posey on Jari Florestal. In *Hearings before the Subcommittee on Foreign Affairs, House of Representatives, 96th Congress, Second Session, May 7, June 19, and September 18, 1980: Tropical deforestation*, pp. 428–433. US Government Printing Office, Washington, D.C.

Post, W. M., W. R. Emanuel, P. J. Zinke, and A. G. Stangenberger. 1982. Soil carbon pools and world life zones. *Nature*, **298**: 156–159.

Proctor, J. 1983. Mineral nutrients in tropical forests. *Progress in Physical Geography*, **7**: 422–431.

Proctor, J., J. M. Anderson, P. Chai, and H. W. Vallack. 1983a. Ecological studies in four contrasting lowland rain forests in Gunung Mulu National Park, Sarawak. I. Forest environment, structure, and floristics. *Journal of Ecology*, **71**: 237–260.

Proctor, J., J. M. Anderson, S. C. L. Fogden, and H. W. Vallack. 1983b. Ecological studies in four contrasting lowland rain forests in Gunung Mulu National Park, Sarawak. II. Litterfall, litter standing crop and preliminary observations on herbivory. *Journal of Ecology*, **71**: 261–283.

Prospero, J. M., R. A. Glaccum, and R. T. Nees. 1981. Atmospheric transport of soil dust from Africa to South America. *Nature*, **289**: 570–572.

Quiroz Flores, A. 1980. Papel de algunas hidrofitas en la fertilidad del sistema chinampero. *Biotica (Mexico)*, **5**: 169–179.

Raich, J. W. 1980a. Carbon budget of a tropical soil under mature wet forest and young vegetation. MS thesis, University of Florida, Gainesville, Fla.

Raich, J. W. 1980b. Fine roots regrow rapidly after forest felling. *Biotropica*, **12**: 231–232.

Reichle, D. E., P. B. Dunaway, and D. J. Nelson. 1970. Turnover and concentration of radionuclides in food chains. *Nuclear Safety*, **11**: 43–55.

Rice, E. L., and S. K. Pancholy. 1972. Inhibition of nitrification by climax ecosystems. *American Journal of Botany*, **59**: 1033–1040.

Richards, P. W. 1952. *The tropical rain forest*. Cambridge University Press, Cambridge.

Richardson, C. J., and J. A. Lund. 1975. Effects of clear cutting on nutrient losses in aspen forests on three soil types in Michigan. In F. G. Howell, J. B. Gentry, and M. H. Smith (eds), *Mineral cycling in southeastern ecosystems*, pp. 673–686. Energy Research and Development Administration, Washington, D.C.

Risch, S. J. 1981. Insect herbivore abundance in tropical monocultures and polycultures: an experimental test of two hypotheses. *Ecology*, **62**: 1325–1340.

Rodin, L. E., and N. I. Bazilevich. 1967. *Production and mineral cycling in terrestrial vegetation*. Oliver and Boyd, Edinburgh.

Ruehle, J. L., and D. H. Marx. 1979. Fiber, food, fuel, and fungal symbionts. *Science*, **206**: 419–422.

Rumney, G. R. 1968. *Climatology and the world's climates*. Macmillan, New York.

Runge, M. 1983. Physiology and ecology of nitrogen nutrition. In O. L. Lange, P. S. Nobel, C. B. Osmond, and H. Ziegler (eds), *Physiological plant ecology, Vol. III, Responses to the Chemical and biological environment*, pp. 163–200. Springer-Verlag, Berlin.

Russell, A. E. 1983. Nutrient leaching during large storms in tropical successional ecosystems. Master's thesis, Universityf Florida, Gainesville, Fla.

Russell, C. E. 1983. Nutrient cycling and productivity in native and plantation forests at Jarí Florestal, Para, Brazil. PhD dissertation, University of Georgia, Athens, Ga.

Sabhasri, S. 1978. Effects of forest fallow cultivation on forest production and soil. In P. Kunstadter, E. C. Chapman, and S. Sabhasri(eds), *Farmers in the forest*, pp. 160–184. University of Hawaii Press, Honolulu.

Sanchez, P. A. 1976. *Properties and management of soils in the tropics*. Wiley, New York.

Sanchez, P. A. 1977. Advances in the management of Oxisols and Ultisols in tropical South America. In *Proceedings of the international seminar on soil environment and fertility management in intensive agriculture*, pp. 535–566. Society of Soil Science and Manure, Tokyo, Japan.

Sanchez, P. A. 1981. Soils of the humid tropics. In *Blowing in the wind: deforestation and long-range implications*, pp. 347–410. Department of Anthropology, College of William and Mary, Williamsburg. Va.

Sanchez, P. A., D. E. Bandy, J. H. Villachica, and J. J. Nicholaides. 1982. Amazon basin soils: management for continuous crop production. *Science*, **216**: 821–827.

Sanchez, P. A., J. H. Villachica, and D. E.Bandy. 1983. Soil fertility dynamics after clearing a tropical rainforest in Peru. *Soil Science Society of America Journal*, **47**: 1171–1178.

Sarmiento, G., and M. Monasterio. 1983. Life forms and phenology. In F. Bourliere (ed.), *Ecosystems of the world*, Vol. 13, *Tropical savannas*, pp. 79–108. Elsevier, Amsterdam.

Savage, J. M., C. R. Goldman, D. P. Janos, A. E. Lugo, P. H. Raven, P. A. Sanchez, and H. G. Wells. 1982. *Ecological aspects of development in the humid tropics*. National Academy Press, Washington, D.C.

Schmidt, R. 1981. Summary of Jarí Plantation data, supplied by Inventory and Control Manager, Jarí Florestal, c/o National Bulk Carriers, 1345 Avenue of the Americas, New York, N.Y.

Scott, G. A. J. 1978. Grassland development in the Gran Pajonal of eastern Peru. PhD dissertation, Department of Geography, University of Hawaii.

Seastedt, T. R., and D. A. Crossley. 1984. The influence of arthropods on ecosystems. *BioScience*, **34**: 157–161.

Sedjo, R. A., and M. Clawson. 1983. How serious is tropical deforestation? *Journal of Forestry*, **81**: 792–794.

Shugart, H. H., D. E. Reichle, N. T. Edwards, and J. R. Kercher. 1976. A model of calcium cycling in an east-Tennessee *Liriodendron* forest: model structure parameters and frequency response analysis. *Ecology*, **57**, 99–109.

Silander, S. 1984. Succession at the El Verde radiation site: a 17 year record. Mimeograph, Center for Energy and Environment Research, Rio Piedras, Puerto Rico.

Sinclair, A. R. E., and M. Norton-Griffiths (eds). 1979. *Serengeti: dynamics of an ecosystem*. University of Chicago, Chicago, Ill.

Singer, R., and I. J. Silva Araujo. 1979. Litter decomposition and ectomycorrhiza in Amazonian forests. *Acta Amazonica*, **9**: 25–41.

Singh, J. S., and S. R. Gupta. 1977. Plant decomposition and soil respiration in terrestrial ecosystems. *Botanical Review*, **43**: 449–528.

Sioli, H. 1973. Recent human activities in the Brazilian Amazon region, and their ecological effects. In B. J. Meggars, E. S. Ayensu, and W. D. Duckworth (eds), *Tropical Forest Ecosystems in Africa and South America: A Comparative Review*. pp. 321–334. Smithsonian Institution Press, Washington, D.C.

Smith, W. H., F. H. Bormann,and G. E. Likens. 1968. Response of chemoautotrophic nitrifiers to forest cutting. *Soil Science*, **106**: 471–473.

Snedaker, S. 1970. Ecological studies on tropical moist forest succession in eastern lowland Guatemala. PhD dissertation. University of Florida, Gainesville, Fla.

Snedaker, S. 1984. Coastal, marine, and aquatic ecology: an overview. In A. H. Westing (ed.), *Herbicides in war, the long-term ecological and human consequences*, pp. 95–107. Taylor and Francis, London.

Sobrado, M. A., and E. Medina. 1980. General morphology, anatomical structure, and nutrient content of sclerophyllous leaves of the 'bana' vegetation of Amazonas. *Oecologia*, **45**: 341–345.

Söderlund, R. 1981. Dry and wet deposition of nitrogen compounds. In F. E. Clark and T. Rosswall (eds), *Terrestrial nitrogen cycles. Ecological Bulletins, (Stockholm)*, **33**: 123–130.

Soermarwoto, O. 1977. Nitrogen in tropical agriculture: Indonesia as a case study. *Ambio*, **6**: 162–165.

Solbrig, O. T., and G. H. Orians. 1977. The adaptive characteristics of desert plants. *American Scientist*, **65**: 412–421.

Sollins, P., K. Cromack, R. Fogel, and C. Yan Li. 1981. Role of low-molecular weight organic acids in the inorganic nutrition of fungi and higher plants. In D. T. Wicklow and G. C. Carroll (eds), *The fungal community, its organization and role in the ecosystem*, pp. 607–620. Marcel Dekker, New York.

Sollins, P., C. C. Grier, F. M. McCorison, K. Cromack and R. Fogel. 1980. The internal element cycles of an old-growth Douglas-fir ecosystem in western Oregon. *Ecological Monographs*, **50**: 261–285.

Sollins, P., and F. M. McCorison. 1981. Nitrogen and carbon solution chemistry of an old-growth coniferous forest watershed before and after cutting. *Water Resources Research*, **17**: 1409–1418.

Sommer, A. 1976. *Attempt at a global appraisal of the tropical moist forests*. Publication PO:FDT/76/4, Food and Agriculture Organization of the United Nations, Rome.

Souza Serrão, E. A., I. C. Falesi, J. Bastos de Veiga, and J. F. Teixeira Neto. 1978. Productivity of cultivated pastures on low fertility soils in the Amazon of Brazil. In P. A. Sanchez and L. E. Tergas (eds), *Pasture production in acid soils of the tropics*, Proceedings of a seminar held at CIAT, Cali, Colombia, April 1978, pp. 195–225. Centro Internacional de Agricultura Tropical, Cali, Colombia.

Sprick, E. G. 1979. Composicion mineral y contenido de fenoles foliares de especies leñosas de tres bosques contrastantes de la region Amazonica. Thesis, Licenciado en Biologia, Universidad Central de Venezuela, Escuela de Biologia.

Sprugel, D. G. 1984. Density, biomass, productivity, and nutrient-cycling changes during stand development in wave-regenerated balsam fir forests. *Ecological Monographs*, **54**: 165–186.

St. John, T. V. 1980. Root size, root hairs and mycorrhizal infection: a re-examination of Baylis's hypothesis with tropical trees. *New Phytologist*, **84**: 483–487.

St. John, T. V., and A. B. Anderson. 1982. A re-examination of plant phenolics as a source of tropical black water rivers. *Tropical Ecology*, **23**: 151–154.

St. John, T. V., and D. C. Coleman. 1983. The role of mycorrhizae in plant ecology. *Canadian Journal of Botany*, **61**: 1005–1014.

Stark, N. 1971. Nutrient cycling. II. Nutrient distribution in Amazonian vegetation. *Tropical Ecology*, **12**: 177–201.

Stark, N. M., and C. F. Jordan. 1978. Nutrient retention by the root mat of an Amazonian rain forest. *Ecology*, **59**: 434–437.

Stevenson, I. L. 1964. Biochemistry of soil. In F. E. Bear (ed.), *Chemistry of the soil*, pp. 242–291. Reinhold, New York.

Stewart, W. D. P. 1967. Nitrogen-fixing plants. *Science*, **158**: 1426–1432.

Stewart, W. D. P. 1977. Present-day nitrogen-fixing plants. *Ambio*, **6**: 166–173.

Stewart, W. D. P., M. J. Sampaio, A. O. Isichei, and R. Sylvester-Bradley. 1978. Nitrogen fixation by soil algae of temperate and tropical soils. In J. Dobereiner, R. H. Burris, and A. Hollaender (eds), *Limitations and potentials for biological nitrogen fixation in the tropics*, pp. 41–73. Plenum, New York.

Stommel, H., and E. Stommel. 1979. The year without a summer. *Scientific American*, **240** (6): 176–186.

Stone, E. L., and R. Kszystyniak. 1977. Conservation of potassium in the *Pinus resinosa* ecosystem. *Science*, **198**: 192–194.

Stothers, R. B. 1984. The great Tambora eruption in 1815 and its aftermath. *Science* **224**: 1191–1198.

Stradling, D. J. 1978. Food and feeding habits of ants. In M. V. Brian (ed.), *Production ecology of ants and termites*, pp. 81–106. Cambridge, University Press, Cambridge.

Stumm, W., and J. J. Morgan. 1981. *Aquatic chemistry*. Wiley, New York.

Swaine, M. D., and J. B. Hall. 1983. Early succession on cleared forest land in Ghana. *Journal of Ecology*, **71**: 601–627.

Swank, W. T. 1984. Atmospheric contributions to forest nutrient cycling. *Water Resources Bulletin*, **20**: 313–322.

Swank, W. T., and J. E. Douglass. 1977. Nutrient budgets for undisturbed and manipulated hardwood forest ecosystems in the mountains of North California. In D. L. Correll (ed.), *Watershed research in eastern North America*, pp. 343–364. Chesapeake Bay Center for Environmental Studies, Edgewater, Md.

Swank, W. T., J. B. Waide, D. A. Crossley, and R. L. Todd. 1981. Insect defoliation enhances nitrate export from forest ecosystems. *Oecologia*, **51**: 297–299.

Swank, W. T., J. W. Fitzgerald, and J. T. Ash. 1984. Microbial transformation of sulfate in forest soils. *Science*, **223**: 182–184.

Swift, M. J., O. W. Heal, and J. M. Anderson. 1979. *Decomposition in terrestrial ecosystems* (Studies in Ecology, Vol. 5). University of California Press, Berkeley, Calif.

Synnott, T. J., and R. H. Kemp. 1976. Choosing the best silvicultural system. *Unasylva*, **28** (112–113): 74–79.

Tanner, E. V. J. 1977. Four montane rain forests of Jamaica: a quantitative characterization of the floristics, the soils, and the foliar mineral levels and a discussion of the interrelations. *Journal of Ecology*, **65**: 883–918.

Tanner, E. V. J. 1980a. Studies on the biomass and productivity in a series of montane rain forests in Jamaica. *Journal of Ecology*, **68**: 573–588.

Tanner, E. V. J. 1980b. Litterfall in montane rain forests of Jamaica and its relation to climate. *Journal of Ecology*, **68**: 833–848.

Taylor, A. W. 1961. Review of the effects of siliceous dressings on the nutrient status of soils. *Agricultural and Food Chemistry*, **9**: 163–165.

Thurston, H. D. 1969. Tropical agriculture: a key to the world food crises. *BioScience*, **19**: 29–34.

Time. 1976. Ludwig's wild Amazon kingdom. *Time*, November 15, 1976: 59–59A.

Time. 1979. Billionaire Ludwig's Brazilian gamble. *Time*, September 10, 1979: 76–78.

Time. 1982. End of a billion-dollar dream. *Time*, January 25, 1982: 59.

Toon, O. B., and J. B. Pollack, 1980. Atmospheric aerosols and climate. *American Scientist*, **68**: 268–277.

Tricart, J. 1972. *The landforms of the humid tropics, forests, and savannas*. Longman, London.

Tschirley, F. H. 1969. Defoliation in Vietnam. *Science*, **163**; 779–786.

Tsutsumi, T. 1971. Accumulation and circulation of nutrient elements in forest ecosystems. In P. Duvigneaud (ed.), *Productivity of forest ecosystems*, Proceedings of the Brussels symposium organized by UNESCO and the International Biological Programme, pp. 543–552. UNESCO, Paris.

Turner, B. L., and P. D. Harrison. 1981. Prehistoric raised-field agriculture in the Maya lowlands. *Science*, **213**: 399–405.

Turner, J., and M. J. Singer. 1976. Nutrient distribution and cycling in a sub-alpine coniferous forest ecosystem. *Journal of Applied Ecology*, **13**: 295–301.

Turvey, N. D. 1974. Water in the nutrient cycle of a Papuan rain forest. *Nature*, **251**: 414–415.

Uehara, G., and G. Gillman. 1981. *The mineralogy, chemistry, and physics of tropical soils with variable charge clays*. Westview Press, Boulder, Colo.

Uhl, C. 1982. Tree dynamics in a species rich tierra firme forest of the Amazon Basin. *Acta Cientifica Venezolana*, **33**: 72–77.

Uhl, C. 1984. Rio Negro forest perturbations: the recovery process. In E. Medina, R. Herrera, C. F. Jordan, H. Klinge, and C. Uhl (eds), *Structure and function of Amazonian forest ecosystems in the upper Rio Negro*. Junk, The Hague, in preparation.

Uhl, C. 1985. Nutrient concentrations and root/shoot ratios in plants occurring on disturbed Amazonian sites. Manuscript in preparation.

Uhl, C., and K. Clark. 1983. Seed ecology of selected Amazon basin successional species. *Botanical Gazette*, **144**: 419–425.

Uhl, C., and C. F. Jordan. 1984. Vegetation and nutrient dynamics during the first five years of succession following forest cutting and burning in the Rio Negro region of Amazonia. *Ecology*, **65**: 1476–1490.

Uhl, C., K. Clark, H. Clark, and P. Murphy. 1981. Early plant succession after cutting and burning in the upper Rio Negro region of the Amazon basin. *Journal of Ecology*, **69**: 631–649.

Uhl, C., H. Clark, and K. Clark. 1982a. Successional patterns associated with slash-and-burn agriculture in the upper Rio Negro region of the Amazon Basin. *Biotropica*, **14**: 249–254.

Uhl, C., C. Jordan, K. Clark, H. Clark, and R. Herrera. 1982b. Ecosystem recovery in Amazon caatinga forest after cutting, cutting and burning, and bulldozer clearing treatments. *Oikos*, **38**: 313–320.

UNESCO. 1978. *Tropical forest ecosystems*, a state-of-knowledge report prepared by UNESCO/UNEP/FAO. United Nations Educational, Scientific, and Cultural Organization, Paris.

Vandermeer, J. 1981. The interference production principle: an ecological theory for agriculture. *BioScience*, **31**: 361–364.

Vandermeer, J. H., S. R. Gliessman, K. Yih, and M. Amador. 1983. Overyielding in a corn-cowpea system in southern Mexico. *Biological Agriculture and Horticulture*, **1**: 83–96.

Van Huay, H. 1984. Soil ecology: symposium summary. In A. H. Westing (ed.), *Herbicides in war, the long-term ecological and human consequences*, pp. 65–67. Taylor and Francis, London.

Van Trung, T. 1984. Terrestrial plant ecology and forestry: symposium summary. In A. H. Westing (ed.), *Herbicides in war, the long-term ecological and human consequences*, pp. 27–29. Taylor and Francis, London.

Van Wambeke, A. 1978. Properties and potentials of soils in the Amazon basin. *Interciencia*, **3**: 233–242.

Verstraete, W. 1981. Nitrification. In F. E. Clark and T. Rosswall (eds), *Terrestrial nitrogen cycles. Ecological Bulletins, (Stockholm)* **33**: 303–314.

Villachica, J. H., and P. A. Sanchez. 1978. Micronutrient research. In *Agronomic-economic research on soils of the tropics: annual report for 1976—1977*, pp. 50–63. Soil Science Department, North Carolina State University, Raleigh, N. C.

Villachica, J. H., C. E. Lopez, and P. A. Sanchez. 1976. Continuous cropping experiment. In *Agronomic-economic research on tropical soils: annual report for 1975*, pp. 117–137. Soil Science Department, North Carolina State University, Raleigh, N. C.

Visser, S. A. 1964. Origin of nitrates in tropical rainwater. *Nature*, **201**: 35–36.

Vitousek, P. 1982. Nutrient cycling and nutrient use efficiency. *American Naturalist*, **119**: 553–572.

Vitousek, P. 1984. Litterfall, nutrient cycling, and nutrient limitation in tropical forests. *Ecology*, **65**: 285–298.

Vitousek, P. M., and P. A. Matson. 1984. Mechanisms of nitrogen retention in forest ecosystems: a field experiment. *Science*, **225**: 51–52.

Vitousek, P. M., J. R. Gosz, C. C. Grier, J. M. Melillo, W. A. Reiners, and R. L. Todd. 1979. Nitrate losses from disturbed ecosystems. *Science*, **204**: 469–474.

Volobuev, V. R. 1964. *Ecology of soils* (translated by Israel Program for Scientific Translations). Daniel Davey and Company, New York.

Wade, M. K., and P. A. Sanchez. 1983. Mulching and green manure applications for continuous crop production in the Amazon Basin. *Agronomy Journal*, **75**: 39–45.

Wadsworth, F. H. 1981. Management of forest lands in the humid tropics under sound ecological principles. In F. Mergen (ed.), *International symposium on tropical forest utilization and conservation*, pp. 168–180. Yale School of Forestry, New Haven, Conn.

Wadsworth, F. H. 1983. Production of usable wood from tropical forests. In F. B. Golley (ed.), *Ecosystems of the World*, Vol. 14A, *Tropical rain forest ecosystems: structure and function*, pp. 279–288. Elsevier, Amsterdam.

Walbridge, M., and P. Vitousek. 1984. Factors relating to phosphorus availability in acid organic coastal plain soils. *Bulletin of the Ecological Society of America*, **65** (2): 236.

Walker, T. W. and J. K. Syers. 1976. The fate of phosphorus during pedogenesis. Geoderma, **15**: 1–19.

Wallace, A. R. 1878. *Tropical Nature and Other Essays*. Macmillan, London; reprinted by AMS Press, New York, 1974.

Walter, H. 1971. *Ecology of tropical and subtropical vegetation*. Oliver and Boyd, Edinburgh.

Waring, R. H., and J. F. Franklin. 1979. Evergreen coniferous forests of the Pacific Northwest. *Science*, **204**: 1380–1385.

Watters, R. F. 1971. Shifting cultivation in Latin America. FAO Forestry Development Paper No. 17, Food and Agriculture Organization, Rome.

Weaver, J. E., and F. E. Clements. 1929. *Plant ecology*. McGraw-Hill, New York.

Weaver, P. 1979. Agri-silviculture in tropical America. *Unasylva*, **31** (126): 2–12.

Weaver, P. L., M. D. Byer, and D. L. Buck. 1973. Transpiration rates in the Luquillo mountains of Puerto Rico. *Biotropica*, **5**: 123–133.

178

Webb, W. L., W. K. Lauenroth, S. R. Szarek, and R. S. Kinerson. 1983. Primary production and abiotic controls in forests, grasslands, and desert ecosystems in the United States. *Ecology*, **64**: 134–151.

Webster, J. R., and B. C. Patten. 1979. Effects of watershed perturbation on stream potassium and calcium dynamics. *Ecological Monographs*, **49**: 51–72.

Went, F. W., and N. Stark. 1968a. Mycorrhiza. *BioScience*, **18**: 1035–1039.

Went, F. W., and N. Stark. 1968b. The biological and mechanical role of soil fungi. *Proceedings of the National Academy of Sciences*, **60**: 497–504.

Werger, M. J. A., and G. A. Ellenbroek. 1978. Leaf size and leaf consistence of a riverine forest formation along a climatic gradient. *Oecologia*, **34**: 297–308.

Werner, P. 1982. The variation in soil properties under tropical rain forest succession in Costa Rica. Unpublished manuscript.

Westing, A. H. 1984. Herbicides in war: past and present. In A. H. Westing (ed.), *Herbicides in war, the long-term ecological and human consequences*, pp. 3–24. Taylor and Francis, London.

Whicker, F. W., and V. Schultz. 1982. *Radioecology: nuclear energy and the environment*, Vol. I. CRC Press, Boca Raton, Fla.

Whitmore, T. C. 1975. *Tropical rain forests of the Far East*. Clarendon Press, Oxford.

Whitmore, T. C. 1978. Gaps in the forest canopy. In P. B. Tomlinson and M. H. Zimmermann (eds), *Tropical trees as living systems*, pp. 639–655. Cambridge University Press, Cambridge.

Whitmore, T. C. 1983. Secondary succession from seed in tropical rain forests. *Commonwealth Forestry Bureau, Forestry Abstracts* **44**: 767–779.

Whittaker, R. H. 1975. *Communities and ecosystems*. Macmillan, New York.

Whittaker, R. H., and P. P. Feeny. 1971. Allelochemics: chemical interactions between species. *Science*, **171**: 757–770.

Whittaker, R. H., and G. E Likens. 1975. The biosphere and man. In H. Lieth and R. H. Whittaker (eds), *Primary productivity of the biosphere* (Ecological Studies Vol. 14), pp. 305–328. Springer-Verlag, New York.

Whittaker, R. H., G. E Likens, F. H. Bormann, J. S. Eaton, and T. G. Siccama. 1979. The Hubbard Brook ecosystem study: forest nutrient cycling and element behavior. *Ecology*, **60**: 203–220.

Whittaker, R. H., and P. L. Marks. 1975. Methods of assessing terrestrial productivity. In H. Lieth and R. H. Whittaker (eds), *Primary productivity of the biosphere* (Ecological Studies, Vol. 14), pp. 55–118. Springer-Verlag, New York.

Wiegert, R. G. 1970. Effects of ionizing radiation on leaf fall, decomposition, and litter microarthropods of a montane rain forest. In H. T. Odum and R. F. Pigeon (eds), *A tropical rain forest*, pp. H-89–H-100. US Atomic Energy Commission, Washington, D.C.

Wiklander, L. 1964. Cation and anion exchange phenomena. In F. E. Bear (ed.), *Chemistry of the soil*, pp. 163–205. Reinhold, New York.

Wiley, G. R. 1982. Maya archaeology. *Science*, **215**: 260–267.

Williams, C. H., and J. D. Colwell. 1977. Inorganic chemical properties. In J. S. Russell, and E. L. Greacen (eds), *Soil factors in crop production in a semi-arid environment*, pp. 105–126. University of Queensland Press, St. Lucia, Queensland.

Williams, J. D. H., J. K. Syers, and T. W. Walker. 1967. Fractionation of soil inorganic phosphate by a modification of Chang and Jackson's procedure. *Soil Science Society of America Proceedings*, **31**: 736–739.

Williams-Linera, G. 1983. Biomass and nutrient content in two successional stages of tropical wet forest in Uxpanapa, Mexico. *Biotropica*, **15**: 275–284.

Witkamp, M. 1970. Mineral retention by epiphyllic organisms. In H. T. Odum and *Nuclear Safety*, **8**: 58–62.

Witcamp, M. 1970. Mineral retention by epiphyllic organisms. In H. T. Odum and R. F. Pigeon (eds), *A tropical rain forest*, pp. H-177–H-179. Division of Technical Information, US Atomic Energy Commission, Washington, D.C.

Witkamp, M. 1971. Soils as components of ecosystems. *Annual review of Ecology and Systematics*, **2**: 85–100.

Woessner, R. A. 1982. Plantation forestry and natural forest utilization in the Amazon Basin. Paper presented at the American Society of Foresters meeting, September 19–22, Cincinnati, Ohio.

Wolfe, J. A. 1978. A paleobotanical interpretation of Tertiary climates in the northern hemisphere. *American Scientist*, **66**: 694–703.

Wolfe, J. N. 1963. Impact of atomic energy on the environment and environmental science. In V. Schultz and A. W. Klement (eds), *Radioecology: proceedings of the first national symposium*, pp. 1–2. Reinhold, New York.

Wood, T. G. 1978. Food and feeding habits of termites. In M. V. Brian (ed.), *Production ecology of ants and termites*, pp. 55–80. Cambridge University Press, Cambridge.

Wood, T. G., and W. A. Sands. 1978. The role of termites in ecosystems. In M. V. Brian (ed.), *Production ecology of ants and termites*, pp. 245–292. Cambridge University Press, Cambridge.

Woodmansee, R. G., and L. S. Wallach. 1981. Effects of fire regimes on biogeochemical cycles. In *Fire regimes and ecosystem properties*, proceedings of a conference in Honolulu, Hawaii, December 1978, pp. 379–400. US Department of Agriculture, Technical Report WO-26, Washington, D. C.

Woodwell, G. M. 1967. Radiation and the patterns of nature. *Science* **156**: 461–470.

Woodwell, G. M. 1970. Effects of pollution on the structure and physiology of ecosystems. *Science*, **168**: 429–432.

Woodwell, G. M., and R. H. Whittaker. 1967. Primary production and the cation budget of the Brookhaven forest. In H. E. Young (ed.), *Symposium on primary productivity and mineral cycling in natural ecosystems*, pp. 151–166. College of Life Sciences and Agriculture, University of Maine, Orono, Maine.

Woodwell, G. M., R. H. Whittaker, and R. A. Houghton. 1975. Nutrient concentrations in plants in the Brookhaven oak–pine forest. *Ecology*, **56**: 318–332.

Wright, R. F. 1976. The impact of forest fire on the nutrient fluxes to small lakes in northeastern Minnesota. *Ecology*, **57**: 649–663.

Yoda, K. 1978. Organic carbon, nitrogen, and mineral nutrients stock in the soils of Pasoh forest. *Malaysian Nature Journal*, **30**: 229–251.

Yoda, K., and T. Kira 1969. Comparative ecological studies on three main types of forest vegetation in Thailand. V. Accumulation and turnover of soil organic matter, with notes on the altitudinal soil sequence on Khao (Mt.) Luang peninsular Thailand. *Nature and Life in Southeast Asia*, **6**: 83–112.

Zinke, P. J. 1984. Soil ecology: an overview. In A. H. Westing (ed.), *Herbicides in war, the long-term ecological and human consequences*, pp. 75–81. Taylor and Francis, London.

Zinke, P. J., S. Sabhasri, and P. Kunstadter. 1978. Soil fertility aspects of the Lua' forest fallow system of shifting cultivation. In P. Kunstadter, E. C. Chapman, and S. Sabhasri (eds), *Farmers in the forest*, pp. 134–159. University of Hawaii, Honolulu.

Zinke, P. J., A. G. Stangenberger, W. M. Post, W. R. Emanuel, and J. S. Olson. 1984. *Worldwide organic soil carbon and nitrogen data*. Oak Ridge National Laboratory, Environmental Sciences Division Publication No. 2212, US Department of Energy.

Subject and Author Index

181

184